ENGINEERED IRONY.

Crossing Octave Chanute's
Kansas City Bridge
for Trains and Teams
1867-1917

Volume 1

by David W. Jackson

150th Commemorative Edition

The Orderly Pack Rat
Kansas City, Missouri
2020

Jackson, David W. *Engineered Irony: Crossing Octave Chanute's Kansas City Bridge for Trains and Teams, 1867-1917*. (Kansas City, Mo.: The Orderly Pack Rat, 2020).

Library of Congress Cataloging in Publication.

Jackson, David W., 1969-
    Engineered Irony: Crossing Octave Chanute's Kansas City Bridge for Trains and Teams, 1867-1917. Volume 1: The Kansas City Bridge.
226 p.   cm.
    Includes bibliographical references, illustrations, and index.

ISBN-13:   978-1-7343686-4-2   (The Orderly Pack Rat)

Library of Congress Control Number: 2021910881

1. Missouri River—History. 2. Kansas City (Mo.)—History. 3. Jackson County (Mo.)—History. 4. Civil Engineering. 5. Missouri—Kansas City. 6. Missouri—Jackson County. 7. Missouri River. 8. Bridges. 9. Railroads—Mo.—Kansas City—History. I. Jackson, David W. 1969-. II. Title.

150th Commemorative Edition of Octave Chanute and George Morison's 1870 Original.

> Entered according to Act of Congress, in the year 1870, by
>
> D. VAN NOSTRAND,
>
> in the Office of the Librarian of Congress, at Washington.

Cover and book design by David W. Jackson, The Orderly Pack Rat.

Published by:
The Orderly Pack Rat
david.jackson@orderlypackrat.com
orderlypackrat.com

VOLUME 1 CONTENTS.

INTRODUCTION 5
　by David W. Jackson

THE KANSAS CITY BRIDGE.

CHAPTER I.
HISTORY OF THE PROJECT 9
CHAPTER I I.
CHARACTER OF THE WORK 19
CHAPTER I I I.
FOUNDATIONS 33
CHAPTER I V.
MASONRY..73
CHAPTER V.
SUPERSTRUCTURE..78
CHAPTER V I.
OUTFIT 93
CHAPTER V I I.
CALCULATED STRENGTH..96
CHAPTER V I I I.
COST OF THE WORK 115

APPENDIX.

A.—EXTRACTS FROM THE CHARTER OF THE KANSAS CITY, GALVESTON,
　　　AND LAKE SUPERIOR R.R. CO. 123

B. —CHARTER OF THE KANSAS CITY BRIDGE CO............................... 125

C. —ACT AMENDING THE CHARTER OF THE KANSAS CITY, GALVESTON,
　　　AND LAKE SUPERIOR R. R. CO. 129

D. —TRAFFIC ON THE KANSAS CITY BRIDGE,
　　　FROM JULY 13, 1869, TO FEBRUARY 28, 1870 131

E. —TABLES RELATING TO PIER NO. 4 133

F. —TABLES OF STRAINS IN THE FIXED SPANS 137

G. —TABLES OF STRAINS IN THE DRAW 139

H. —LIST OF PERSONS EMPLOYED 141

ERRATA 143

ILLUSTRATIONS: ELEVATION PLATES (REDUCED FROM OVERSIZED FOLDOUTS) 145

150TH COMMEMORATIVE CONTENT.

OCTAVE CHANUTE: "FATHER OF AVIATION" 195
   by Bill Nicks

2020 KANSAS CITY BRIDGE ENGINEERING RETROSPECTIVE 209
   by Bryan Snyder, PE

ABOUT THE AUTHORS 223

VOLUME 1 NOTES..225

INTRODUCTION.

Over the last 20 years I have researched, written, and published on Kansas City and Jackson County, Missouri's rich history, lectured on our diverse heritage, and advocated on behalf of historic preservation opportunities. Along the way, I have added on my own or in concert with others nearly 40 titles to local history bookshelves (orderlypackrat.com).

The breadth and depth of this work product illuminates two-time Pulitzer Prize winning historian David McCullough's declaration:

> *"I cannot think of another landscape of similar size that has had so much American history as Jackson County, Missouri."*
> —David McCullough
> while researching his 1993 biography, *Truman*

McCullough's truism kindles a plethora of events in local history that, indeed, have national import. *Some* of those seminal moments and notable figures include:

* the Spanish and French explorations through native grasslands to Missouri's savannah;

* the epic Westward Movement (trail, steamboat, and rail) experience that involved the Lewis and Clark, Santa Fe, Mormon, Oregon, California, and Pony Express Trails;

* a tumultuous, early Mormon history rife with expulsion, tarring, and feathering;

* a cultural landscape that involved the enslavement of people of African descent;

* sensational stories involving frontier justice with connections to Jesse James and contemporaries in the days before Kansas City shook its 'cow town' image;

* a major stock raising and grain futures trade rivaling Chicago for Grade-A agribusiness;

* divisive, escalating border guerilla warfare preceding the Civil War by seven years;

* the 1861 American Civil War when martial law was first imposed upon U.S. civilians;

* a vibrant fashion design and garment manufacturing beacon, ranking second only to NYC;

* a renowned parks and boulevards system that was emulated by cities across the nation;

* the birthplace of Mickey Mouse, and other Hollywood and Broadway stars: Robert Altman, Ed Asner, Don Cheadle, Peter DeLaurier, Jean Harlow (Harlean Carpenter), Virginia "Ginger" Rogers, Ellie (Elizabeth Claire) Kemper, Diane Wiest, and others;

* Harry S Truman's ascent to the highest office in the land . . . as mob- and boss-politics in Kansas City were swept clean;

* Kansas City's 'women of independent minds' who led the charge on so many local, regional, and national programs that generations have enjoyed and benefitted;

* a tradition in compassion caring for citizens of all ages in need whether it be world class medical, nursing, or hospice services . . . for educating and training medical and dental professionals . . . and for leading biomedical scientific research;

* unnumbered inventions including the shopping cart, multiplex movie theaters, Teflon™, bumper stickers, electric hair clippers, Eskimo Bars, wrapping paper and greeting cards (Hallmark), M&M's candy-coating process, the automatic fire alarm, electric hair clippers, Bomb Pops, and McDonald's Happy Meal™;

* Kansas City pioneer activists who in February 1966 lit the kindling leading to the Gay Liberation Movement sparked at the Stonewall Inn in NYC in June 1969;

* a spectrum of entrepreneurs from Hiram Young and "Russell Stover" (i.e. Clara Stover); and two, true Kansas City *originals:* KC Jazz-Charlie Parker & KC BBQ-Henry Perry;

* a series of sports teams who have scored high across the nation (from the Kansas City Athletics and Monarchs, to the Blades, Chiefs, Comets, Mavericks, Royals, Scouts, T-Bones, and even Sporting Kansas City); and;

* as we, at the doorstep of the Solar Age, barrel towards doomsday climate scientific projections, Kansas City is leading the nation with messaging, projects, and strategies supporting Green New Deal-inspired *ideals* and *similar campaigns* to wean U.S. off gasoline and to champion renewables that might reverse the global warming crisis.

Kansas City and Jackson County's origin stories are among my favorites. These genesis experiences are where the Heart of America truly beats. Here, seeming happenstance, disparate actions and diverse people's actions have formed and continue to shape the unique metropolis that is Kansas City.

INTRODUCTION.

Engineered Irony commemorates the 150th anniversary of one of those momentous Kansas City achievements with national significance that solidified Kansas City's "destiny as the metropolis of the Missouri Valley."[1] Arguably, if this one event had not unfolded as it did, history would have forged a vastly different historical timeline for the region.

In 1870, engineers Octave Alexandre Chanute and George Shattuck Morison published their book, *The Kansas City Bridge*, a report recapping the previous year's completion of their greatest achievement up to that moment.[2]

Chanute and Morison's 1869 "Kansas City Bridge" masterwork—most commonly known as "The Hannibal Bridge" named after the Hannibal & St. Joseph Railroad that was responsible for its creation—was the first, permanent bridge to span the winding, often treacherous, muddy Missouri River. *Here the names are used interchangeably.* That railroaders in Omaha, Nebraska, had laid rails atop the frozen river, drove temporary piles into the ice, and inched a locomotive across the solid glaze in the winters of 1866-71 does not discount Kansas City's auspicious distinction from the standpoint of permanence.[3]

The Hannibal Bridge actually predates the Brooklyn Bridge in New York and the Eads Bridge in St. Louis, so Chanute's experiences likely factored into the planning of those landmark structures.[4] Chanute's modern engineering achievement in iron was accomplished "right here in river city" in an 'ironic' twist of fate.

"President Lincoln picked a route from Omaha for the first rail line to reach the West Coast."[5] The famed "Golden Spike Ceremony" drove the celebrated "last spike" on May 10, 1869, at Promontory Summit in Utah Territory. Technically, that event conjoined the railroad system only from San Francisco, California, to Omaha, Nebraska. The nation's railroad system was not yet *truly* transcontinental until the Missouri River was bridged. At Omaha, that did not happen until March 1872.

Two and a half years *earlier* on July 3, 1869, Kansas City opened its "Hannibal Bridge" with a fete that attracted a conservative estimate of 30,000 spectators. The "last spike" celebrated that day set an unparalleled trajectory for any midwestern city to date. Not only did Kansas City become the hub for coast-to-coast transportation, but it shortened the trip from roughly six months on foot leading an ox-pulled wagon to as little as eight days by train from New York City to San Francisco.[6]

Volume one of this 150th Commemorative Edition reprints Chanute and Morison's 1870 report, *The Kansas City Bridge* published by D. Van Nostrand, with testimonials to Chanute and his engineering genius.[7] Volume two represents my decade plus effort to relive that auspicious day in 1869, and to memorialize the promoters, campaigners, and skilled workers who engineered, constructed, then operated the Hannibal Bridge over the next 48 years . . . and to name the handful who lost their lives in the process.

David W. Jackson
Kansas City, Missouri
December 2020

THE

KANSAS CITY BRIDGE,

WITH AN ACCOUNT OF THE REGIMEN OF THE MISSOURI RIVER,

AND A

DESCRIPTION OF METHODS USED FOR FOUNDING IN THAT RIVER.

BY

O. CHANUTE, Chief Engineer,

AND

GEORGE MORISON, Assistant Engineer.

ILLUSTRATED.

NEW YORK:

D. VAN NOSTRAND, PUBLISHER,

23 Murray Street and 27 Warren Street.

1870.

CHAPTER I.

HISTORY OF THE PROJECT.

THE Missouri River has long been known as so turbulent and unstable a stream, that it was considered by many of those best acquainted with its character, as almost incapable of being bridged. The successful completion of the first bridge across this river, and the novelty of some of the methods adopted for putting down its foundations, especially that introduced at Pier No. 4, which is believed to be capable of considerable extension in similar works, have therefore excited such general attention and inquiry, as to make it seem desirable that some record should be published of its construction.

It is admitted that many of the plans were very different from those which, in the light of present experience, it would be wished to adopt; but it is believed that a narrative of the difficulties and temporary failures on this pioneer work, may prove more interesting and instructive than would be the account of the more matured plans of a second undertaking.

The movement which led ultimately to the building of the Kansas City bridge, dates from the incorporation of the "Kansas City, Galveston, and Lake Superior Railroad" by the State of Missouri, in 1857. This high-sounding title, and the extent of the enterprise, which contemplated some 1,500 miles of railroad, occasioned a great deal of merriment in the Legislature, especially as but 129 miles of the scheme lay in Missouri, within the jurisdiction of the body granting the charter, and it was also understood that the projectors would, for the present, be satisfied with the building of 52 miles of the line, as a branch of another railroad.

But the enterprising citizens of the then infant City of Kansas, which perhaps contained at that time some 2,000 inhabitants, proved wiser than those who laughed at their plans, and they now have the satisfaction of seeing substantially the same road in the course of rapid execution from the Gulf to the great lakes.

In 1860 a contract was let for building that portion of the road extending from the town of Cameron, on the Hannibal and St. Joseph Railroad, to the Missouri River, opposite Kansas City.

Although one provision of this charter, extracts from which will be found in Appendix A, authorized the bridging of all navigable streams within the State (the Missouri being the only river to which this clause could possibly apply on the line above mentioned), yet it was considered so formidable an undertaking, that no steps whatever were taken towards building a bridge, and the line was located parallel with the north bank of the river, through Harlem, a village opposite Kansas City, thence extending in a north-easterly direction.

A good deal of work was done, and some $200,000 expended ; but the breaking out of the civil war put an end to all active operations in the spring of 1861, and for the next five years the project slumbered forgotten in the strife which desolated the border between Missouri and Kansas.

In the year 1865 a charter was obtained from the Legislature of Missouri, for a carriage and railroad bridge at Kansas City, a copy of which will be found in Appendix B. This movement, however, was mainly speculative, and the corporators, having failed to interest the necessary capital, never organized under it, and merely held the charter as a ready means of benefiting the town by giving it to any parties willing to undertake the construction of the bridge, should circumstances ever render such an undertaking probable.

In the following year, the Kansas City, Galveston, and Lake Superior Railroad, which had now been revived, and whose name was at about the same time changed to the "Kansas City and Cameron Railroad," had its charter amended so as to invest it with like privileges as to bridging the Missouri, to those belonging to the Kansas City Bridge Company.

A general Act of Congress was approved on the 25th day of July, 1866, authorizing the construction of bridges across the Mississippi River at Quincy, Burlington, Hannibal, Prairie du Chien, Keokuk, Winona, Dubuque, and St. Louis, which by a special clause was made to apply to the Missouri River at Kansas City.

The restrictions and conditions of the bridges becoming post routes (except for the St. Louis bridge), and the clause applying to the Missouri, were as follows :

*   *   *   *   *   *   *   *   *   *   *   *

SEC. 2. *And be it further enacted*, that any bridge built under the provisions of this Act, may, at the option of the company building the same, be built as a drawbridge, with a pivot or other form of draw, or with unbroken or continuous spans ; *provided*, that if the said bridge shall be made with unbroken and continuous spans, it shall not be of less elevation, in any case, than 50 feet above extreme high-water mark, as understood at the point of location, to the bottom chord of the bridge ; nor shall the spans of said bridge be less than 250 feet in length, and the piers of said bridge shall be parallel with the current of the river ; and the main span shall be over the main channel of the river, and not less than 300 feet in length ; *and provided also*, that if any bridge built under this Act, shall be constructed as a drawbridge, the same shall be constructed as a pivot drawbridge, with a draw over the main channel of the river at an accessible and navigable point, and with spans of not less than 160 feet in length in the clear on each side of the central or pivot pier of the draw, and the next adjoining spans to the draw shall not be less than 250 feet ; and said spans shall not be less than 30 feet above low-water mark, and not less than 10 above extreme high-water mark, measuring to the bottom chord of the bridge, and the piers of said bridge shall be parallel with the current of the river ; *and provided also*, that said draw shall be opened promptly, upon reasonable signal, for the passage of boats whose construction shall not be such as to admit of their passage under the permanent spans of said bridge, except when trains are passing over the same ; but in no case shall unnecessary delay occur in opening the said draw, during or after the passage of trains.

SEC. 3. *And be it further enacted*, that any bridge constructed under this Act, and according to its limitations, shall be a lawful structure, and shall be recognized and known as a post route ; upon which, also, no higher charge shall be made for the transmission over the same, of the mails, the troops, and the munitions of war of the United States, than the rate per mile paid for their transportation over the railroads or public highways leading to the said bridge.

*   *   *   *   *   *   *   *   *   *   *   *

Sec. 10. *And be it further enacted*, that any company authorized by the Legislature of Missouri, may construct a bridge across the Missouri River, at the City of Kansas, upon the same terms and conditions provided for in this Act.

*   *   *   *   *   *   *   *   *   *   *   *

It was early decided, that the alternative clause authorizing the construction of a pivot drawbridge, would be the proper one to adopt for Kansas City. The topography of the river and its banks is such as to confine the choice of a point of crossing opposite the town within narrow limits; while just above, between the Bluffs and the Kansas River, spreads out a flat bottom land, the natural point of connection and exchange between all the railroads centring at this city, occupied by them at an early day. A high bridge crossing would have made it impossible to reach depot grounds, or connect with the other roads on this bottom, without the use of gradients of 90 or 100 feet per mile, and a drawbridge, requiring an elevation of only 10 feet above high-water mark, was accordingly selected.

The provisions of the Act of Congress, concerning drawbridges, were mainly framed to apply to the Mississippi River, and when extended to the Missouri, some of them could seldom be safely complied with to the letter. Thus the requirement that the spans adjoining the draw should be 250 feet each, designed to accommodate the immense rafts which float down the comparatively tranquil current of the Mississippi, becomes useless in the Missouri, whose turbulent torrent forbids the handling of any rafts, save those composed of a few cotton-wood logs, run down along the shore a few miles to the nearest saw-mill. Besides, as at almost every point where a bridge would be likely to be attempted, the channel of the Missouri lies close to one of its shores, the attempt to place spans of 250 feet on each side of the draw would result either in locating one leg of the draw beyond the main channel, or in building one of the 250 feet spans partly over dry land.

This alternative was presented at Kansas City, and it was deemed that the placing of the draw in the best possible location over the main navigable channel, was the chief requirement, and the adjoining spans were arranged, as hereafter stated, to conform to the local circumstances.

The length of the spans of the draw, which was fixed by the Act, was also intended to meet the requirements of the large tows, composed of steamboats and barges, which ply on the Mississippi. But on the Missouri River there are no tows, the rapidity of the current, and the many snags to be found in the stream, rendering the towing of barges hazardous, and having thus far caused the failure of every attempt to introduce them, so that a narrower draw might have answered the requirements of the boats navigating this stream, which are moreover smaller than those running on the Mississippi; yet in view of the extreme swiftness of the current during floods, and of the difficulty of holding a boat at such time perfectly true to her course, this requirement of the law may be deemed a prudent one, and the spans of the draw of the Kansas City bridge were accordingly made each a little over 160 feet in length in the clear.

In 1866, the Kansas City and Cameron Railroad being fully reorganized, with Mr. C. E. Kearney, of Kansas City, as president, obtained additional subscriptions and set about to seek aid and a connection with the Hannibal and St. Joseph Railroad. A curious accident, which occurred in connection with this road, showed on how slender a thread sometimes hangs the fate of infant projects and communities. Even before the war, a strong rivalry existed between Kansas City and Leavenworth, the latter city being located on the same bank some 25 miles up the river. Both had begun railroads to Cameron, both had temporarily abandoned their enterprise during the war, and both sought the aid of the Eastern capitalists controlling the Hannibal and St. Joseph Railroad, to revive them. Leavenworth, which had enjoyed a large and prosperous trade during the war, in consequence of being near an important military post and fort, was earliest in the field, and when Kansas City heard of it, had all but closed a contract for the necessary aid with the Eastern capitalists. A very few days more and it would have been too late; every thing would have been arranged, and the road and bridge built to Leavenworth, which city would probably have been enabled completely to crush her rival. Immediate personal appeals and propositions brought about a suspension of a final judgment, until the claims and merits of the two schemes could be investigated.

This was done by Mr. James F. Joy, himself, who as president or chief manager of the Michigan Central, Chicago Burlington and Quincy, Hannibal and St. Joseph, and other roads, was at that time preparing to have bridges built across the Mississippi River both at Burlington and at Quincy. He visited personally Leavenworth and Kansas City, decided that the latter was the best point to reach, and that a bridge must also be built to make the road of value. Arrangements were therefore entered into between the Kansas City and Cameron and the Hannibal and St. Joseph Railroad companies, by which the capital interested in the latter and connecting lines of railroads, agreed to furnish the iron and equivalent for the new line, and to build the bridge at Kansas City. The entire property of the Kansas City and Cameron Railroad was assigned to Messrs. James F. Joy, Nathaniel Thayer, and Sidney Bartlett, as trustees, and the work was carried on to completion under the fiduciary charge of these gentlemen, Mr. Joy remaining throughout the chief manager of the enterprise.

On the 30th of November, 1867, the railroad was completed from Cameron to the north bank of the river opposite Kansas City, and from that date until the completion of the bridge in July, 1869, the road was operated, as a branch of the Hannibal and St. Joseph Railroad, freight and passengers being transferred by ferry.

A preliminary survey and report on the bridge site had been made in August, 1866, by Mr. M. Hjortsberg, Chief Engineer of the Chicago, Burlington, and Quincy Railroad, and on the 7th day of February following, Mr. Chanute took charge of the work as chief engineer of the bridge, under an appointment from Mr. Joy, and from that time until its completion the work was steadily prosecuted. Owing to the novelty of the work, and the difficult nature of the foundations, no trustworthy contracts could be let for them at that time, and it was determined that the company should do the subaqueous part of the work itself. Pile-driving was begun on the north bank of the river on the 27th of February, but early in April, operations at the bridge site were interrupted by high water, and could not be resumed before the 1st of August.

These spring and summer months were not wasted, but used to advantage in the preparation of a suitable outfit, and in building upon the shores the caissons and cribs afterwards used in the construction of Piers 1 and 2.

Kansas City at that time was almost on the frontier; there was but one small foundry and machine shop in the town, while not a barge suitable to carry stone could be found on the river. Special tools had also to be designed and erected, which, however simple and obvious they may seem now, caused the engineer no little thought and anxiety. A steamboat was also found necessary to tow the barges, and for this purpose the steamer "Gipsey" was purchased in Wheeling, and brought to the bridge site; eight flat-boats were built, and two small ones purchased; pile drivers, derricks, and dredges built, and a dismantled building, situated near the bank of the river, half a mile above the bridge, was bought and converted into a machine shop.

The contract for the masonry was let on the 23d of February, 1867, to Messrs. Vipond & Walker, of Kansas City, and the quarrying of stone was begun forthwith. The corner-stone of the south abutment was laid on the 21st of August, 1867, with appropriate festivities, and the last stone was laid on the 5th of May, 1869, when the completion of Pier No. 4 finished the masonry of the bridge.

A contract for the superstructure was closed with the Keystone Bridge Company, of Pittsburg, on the 22d of November, 1867, and under the direction of that company this portion of the work was carried to completion.

The timber used in the false works, and in the trestle for the northern approach, was mostly native oak lumber, and obtained in small contracts from time to time, whenever needed. Contracts for the grading of the southern approach, and for several unimportant parts of the work, were let during its progress; but, with the exception of the masonry and superstructure, the bulk of the work was all done by the Company.

The draw was swung on the 15th day of June, 1869, the first engine crossing the bridge ten days thereafter, and the bridge was publicly opened on Saturday, July 3d, 1869.

The period of two years and a half, thus consumed in the location and construction of this bridge, had brought about great changes in its immediate vicinity. The population of Kansas City had increased from 13,000 to 30,000, and from being little more than a way-station on the Missouri Pacific Railroad, it had become an important railway centre, from which no less than seven lines of railroad were in full operation, while several more were projected. Though the bridge was originally built only for the use of the Kansas City and Cameron Railroad, seven months before its completion the west branch of the North Missouri Railroad had been finished to Harlem, and this company had made arrangements to run over the bridge, while the Missouri Valley Railroad had been extended from its former terminus opposite Leavenworth to the same point, so that the bridge became at once, not only a link in the line of railroads extending from Chicago to the South-west, but united the railway system of Northern and Southern Missouri and Kansas at a common point, near the boundary of the two States.

On the south bank of the river, the Kansas Pacific Railway, starting from the State line at Kansas City, had been completed 405 miles, nearly to the eastern boundary of Colorado. The Missouri River, Ft. Scott, and Gulf Railroad, a line intended to occupy a part of the ground embraced in the original scheme of the Kansas City, Galveston, and Lake Superior Railroad, had come under the management of the same interests which built the Kansas City and Cameron Railroad and the bridge, was already in operation for about 50 miles, and has since been extended to the south line of Kansas. These with the Pacific Railroad of Missouri, and the Missouri River Railroad, which were operated as one line from St. Louis to Leavenworth, made up the list of seven railroads in operation to Kansas City, while steps were being taken, and subscriptions obtained, for another eastern outlet, to connect with the lines controlled by the Pennsylvania Railroad, by way of the town of Louisiana on the Mississippi River ; for the Kansas City, Springfield and Memphis Railroad, and for the Kansas City and Santa Fe Railroad, a line designed to tap the business of the Leavenworth, Lawrence, and Galveston Railroad, and to extend into South-western Kansas, which is now in process of construction.

The completion of the bridge united the three railroads on the north and east side of the river with the four on the south and west, made Kansas City the convenient point of exchange for all business going south-westerly, and gave it such commercial importance as wellnigh to justify the boast of its sanguine citizens, that it was destined to become the metropolis of the South-west.

It had been the wish of the people of Kansas City that a separate carriageway should be connected with the bridge, and in order to secure this they were prepared to make some concessions. As it would have been desirable to carry both roads at the same level, the cost of both the masonry and super-structure would have been considerably increased thereby; it was, however, considered that as the bridge was but 1,400 feet long, and could be crossed in five minutes by a team, or in two minutes by a train, no very great trouble need be apprehended, with proper police regulations at the ends, in admitting each traffic alternately upon one floor, and it was accordingly decided to lay down a Nicholson pavement upon the bridge, which was in con-sequence made 18 feet wide in the clear, and to throw it open to carriages at all times, except when trains were to pass. All the foundations were, moreover, put in for a double-track bridge, so that it can be widened whenever the traffic becomes so great as to require it. This arrangement, which was not adopted without some misgivings, has thus far been found to work perfectly well. Not the slightest accident has occurred in consequence; and the delay to trains or teams, from finding the bridge occupied upon approaching it by each other, has proved trifling and unimportant. Some idea of the large business done over the bridge, from the very beginning, will be formed by examining the abstract of its traffic for the first seven and a-half months, given in Appendix D.

The chief anxiety of all parties concerned in this work was so to locate and build the bridge that it should form the least possible obstruction to the naviga-tion of the river, and prove as little objectionable as possible to the steamboat interests. It was felt that, whatever other mistakes might be made, the channel must be kept clear, and boats be enabled to pass and repass at all times, with-out danger or difficulty. This was the intention. It is hoped and believed that it has been fulfilled, for thus far not the slightest accident has occurred to any boat in passing the bridge. For two years steamers of all sizes have gone up and down, at all stages of the water, passing the piers and works in progress. After the completion of the bridge, at the suggestion of some gentlemen con-nected with the steamboat interests, a timber dock or shore was provided above the bridge, by swinging a series of pontoons above and in line with the southern pier, to enable boats to drop down along their sides in very dark or windy weather; but although this was only completed and put in place late in the

Engineered Irony

season of 1869, boats passed through all that year without material trouble or delay. This gratifying result must, in great part, be attributed to the care, reason, and justice of the men navigating this river, and has happily avoided the disputes and accidents which have attended the erection of the first bridges across some other of our large rivers.

CHAPTER II.

CHARACTER OF THE WORK.

THE circumstance which gave the most interest to this work, was the fact of its being the pioneer bridge across the Missouri River, and to the distinguishing features of that river the chief difficulties of the undertaking were due.

Of the three great tributaries of the Lower Mississippi, the Missouri is at once the largest, the wildest, and the least known. The Ohio, draining the eastern slope of the Mississippi basin, flowing through a well-settled country, between high banks, over a hard and undisturbed bed, has long proved a most serviceable stream for navigation, and offers no peculiar difficulties to the bridge-builder. The Upper Mississippi, rising among the plains of the central valley, and flowing for its whole navigable length through a low bottom land between the high bluffs which mark the level of the surrounding country, has in general a sandy and somewhat unstable bed; but its light fall and easy current render it a good river to navigate, while its regimen is sufficiently fixed to make the task of bridging more properly one of magnitude than of special difficulty. The Missouri, drawing its source from the eastern face of the Rocky Mountains, and flowing with a rapid descent down the westerly slope of the great basin, unites within itself all elements of unstableness and irregularity, combining the impetuosity of a mountain torrent with the volume of a lowland river. The navigable length from Fort Benton to its junction with the Mississippi, is computed by the river pilots at about 3,150 miles, and the area of its drainage is given by Humphreys and Abbot as 518,000 square miles, or more than one-third greater than the united basins of the Ohio and Upper Mississippi. Owing to the lightness of the rain-fall on a large part of this district, the mean annual discharge is far from being proportionate to the extent of the drainage, and the overwhelming floods of exceptional years must be taken as the real examples of the size of the river ; but its greatness is also shown by the character of the water, filled with a light sand brought from the disintegrating rocks among the mountains, by the strange geological mixture found in the gravel and pebbles below its bed, and by the annual summer floods which come in their greatest violence when other rivers are on the decline.

The chief tributaries of the Missouri are the Yellowstone, the Platte, the Kaw or Kansas, and the Osage—the two latter being prairie streams of irregular supply, and the two former, like the upper river, deriving most of their water from the mountains. Each of these rivers has its own characteristics and produces its distinctive freshets. The Yellowstone unites with the upper river to cause the summer flood ; the Platte usually pours out its water a little earlier in the spring ; while the freshets of the Kaw and the Osage are of less regular occurrence, and dependent largely on local rains. A combination of these freshets, the waters from the melting snows among the mountains being supplemented by heavy rains in the lower countries, has produced the great floods which occur at long but irregular intervals, the last of which took place in 1844. This flood, the only great flood of which we have accurate information, submerged the entire bottom land below the mouth of the Kaw, and has been regarded by the settlers as an event too terrible to occur a second time ; but Indian traditions mention other floods of similar character, one of which, occurring towards the end of the eighteenth century, probably in 1785, is said to have considerably exceeded that of 1844.

In many matters of topography the Missouri resembles the Upper Mississippi, while it is substantially identical in these respects with the lower river. Its course lies through a low alluvial deposit of bottom land enclosed on each side by bluffs. The distance between these bluffs varies from a mile and a half to 15 miles or more, the bluffs being generally highest, most rugged, and containing the greatest quantity of rock, when they approach most nearly together. For about 500 miles from the mouth of the river, or nearly to the southern boundary of the State of Nebraska, the bottom land, except where artificially cleared or where its width is very great, is covered with a heavy growth of timber, the cotton-wood being the most common tree, while the sycamore, black walnut, and several varieties of oak and elm also abound ; farther north the timber becomes more scarce, and a large part of the bottom land is open prairie. The average elevation of the bottoms is a few feet above the ordinary high-water level, but below the range of the extreme floods of exceptional years. The river winds to and fro in a circuitous course between the bluffs, with little apparent regularity—the width from bank to bank, measured between the wooded or grass-grown shores, varying from 300 to 1,500 yards, and averaging about half a mile. At low water the channel contracts within much smaller limits, becoming reduced to 600 or 700 feet, and leaving the remaining width a dry and desolate sand bar.

The usual fall being from 10 inches to a foot in the mile, the current is very rapid, varying with the different stages of water, in an ordinary season, from three miles an hour to eight. The bed of the river, the sand bars and the substratum of the bottom lands are composed partially of sand and partially of a fine silt, having a specific gravity little greater than that of water* ; a considerable quantity of this silt is always held in suspension by the water, and the current, when strong, moves the combined silt and sand with surprising rapidity. The current is most violent during a rise in the river, and the velocity is dependent on the suddenness of the rise, the level of the water being raised from above, and the surface slope thereby temporarily increased. On these occasions the current is often strong enough to deepen the channel 20 feet in a single day, and if impinging on a low bank, to cut several yards into the shore in a single hour. There is a local saying that the Missouri has a standing mortgage on the entire bottom land from bluff to bluff, and the farmer on the Missouri bottom often learns to his sorrow, by the loss of his farm, that real estate is not always immovable property.

The water of the Missouri is found by analysis to contain less solid matter in solution than is found in the water of any other important river of the continent ; but it always holds a large amount of silt and fine sand in suspen-

---

* The weight of one cubic foot of different varieties of Missouri River sand and silt was found to be as follows :

Coarse sand, dry, 108 lbs., saturated, 132 lbs.
Fine sand, " 101 lbs., " 125 lbs.
Silt, " 87 lbs., " 110 lbs.
Silt, very fine, " 77 lbs.

sion, which, originally emanating from the Yellowstone and the upper river, is from time to time deposited on the bars and again picked up, till it has finally been carried the whole length of the stream and left to form the bars and delta below New Orleans. Portions of the deposits remain undisturbed for centuries, forming the foundation of timber land and perhaps farms ; but many of them are of the most temporary nature, swept away and replaced several times in a season. The character of the deposit also varies materially with different floods ; sometimes it is almost entirely a clayey silt, while at others, especially if the flood be a violent one, it is largely composed of heavy sand. Below the silt and sand there is found a layer of coarse gravel and loose stones of varied geological character, and containing occasional relics of animal life. This gravel deposit is a collection of the coarser portions of the annual flood deposits, which, from its greater weight, having been moved but slowly by the current, has in time settled to the lowest limit of scour ; it is only found at considerable depths, and is almost entirely wanting in those parts of the stream where the bed rock is frequently swept bare.

The water is most nearly clear during the low water of the winter, and especially when the river is frozen ; it is muddiest during the summer flood, when a thickness of half an inch of water becomes a perfectly opaque screen. Such an amount of solid matter can only be kept suspended when in rapid motion, and is at once deposited wherever the current slackens ; hence it usually happens that, while the river is cutting away the bottom lands on one side around a bend, a sand bar is forming on the other ; and after the flood the channel will be found to have changed its position, while its width remains nearly the same as before. The very violence and power of the river thus confine it between narrow banks, and become the masks of its real size, quickly converting any slack-water into dry land, limiting the width to that actually required for the discharge, and depriving the Missouri of such large areas of calm still water as those which add so much to the beauty and apparent size of the Upper Mississippi. For the same reasons the "travelling sands" usually observable in rivers with sandy bottoms, and which have been described at length by writers on the Mississippi River, play a much less active part in the Missouri, as their existence demands a sufficient width of river to allow the whole discharge of water to pass in a shallow stream over the crest of the sand bar ; they undoubtedly occur in this river, but are confined to straight reaches of the stream, where the channel is broad and but poorly defined, and to seasons of high water ; while their action is slow and unimportant compared with the violent wash and scour on the curves or where the current is rapid. When the river cuts into a timbered part of the bottom land the destruction of the bank lets the trees fall into the water ; they usually remain for a short time, seldom as much as a day, at the spot where they fall, forming a temporary protection to the bank and causing local irregularities in the channel ; but they soon become free and float down the stream till caught by some obstruction in the bottom; here they soon lose their leaves and smaller branches, and unless set free by the rising water or by the loosening of the obstruction which entangles them, they remain fast and form the snags which so greatly impede navigation. Besides the live trees washed into the river, every flood picks up a large quantity of loose timber and rubbish from the sand bars and low portions of the bottom land, the amount of drift which even a moderate flood brings down being very great.

When the width of the bottom land is not more than two or three miles, the usual course of the river is to follow along the base of one of the bluffs till deflected by some obstacle, then to cross the valley to the other bluff, follow that for a short distance, and then return to the former side, thus pursuing a serpentine course, and alternately impinging upon each bluff. The meanderings of the river are then more marked and regular than in other streams ; the vein of strongest current can generally be distinguished by a casual observer, it crosses the stream diagonally in the straight reach between the curves, and is always strongest on the outside of the curves ; the river constantly washes the lower bank as it crosses the bottom land, and thereby impinges on the opposite bluff at a lower point than hitherto, so that unless held by natural projections in the bluffs, or other protection, the pair of reversed curves, resembling a large letter S somewhat flattened, and corresponding to the points where the stream leaves one bluff and where it strikes the other, by cutting away the bottom land and forming fresh sand bars, are continually advancing down the valley. When the bottom land has a greater width than two or three miles, the river is liable to turn back to the bluff it has left, before reaching the opposite one ; its course is in this case very irregular, and for more than 100 miles it will sometimes fail to cross the bottom land. Such is the case where the Missouri forms the boundary line between Iowa and Nebraska, the river keeping near the Nebraska bluff almost the entire distance.

The most favorable location for a bridge is just below one of the great bends, especially if the current of the river there impinges upon a rocky shore. Under these circumstances the bed rock, to which the foundations of the channel piers must in general be carried, is found at a comparatively small depth on the bluff side, while the piers on the opposite sand bar can often be founded safely without going to any very great depth ; a moderate stone protection above the bridge will also suffice to secure permanency of channel. The least desirable location is on a long straight reach, especially if bordered on both sides by the low alluvial banks of the bottom land. The bed rock will then usually be found only at great depth, and the current veins are very variable, making it necessary to found all piers at the full depth, and largely increasing the expense and complications of shore protections, as well as requiring a greater length of bridge.

Though the current is too strong to allow the Missouri to freeze directly across, the ice forms rapidly along the banks in cold weather, and a single frosty night will suffice to fill the river with loose cakes of soft ice, which have broken off from the shores. If the weather continues cold, these cakes, rounded by constant attrition and increasing in strength and thickness as they float, freeze to each other, and, finally jamming at some narrow or obstructed point, pack together, and close entirely across, sometimes gorging to the bottom of the river. At Kansas City the river usually closes in this way in December, and continues closed till February, when the ice is sometimes found to be as much as two feet thick ; there is, however, little regularity about it, the river having, in some winters, remained open through the whole season, and in others closing and breaking up several successive times. The river is always low when frozen, and if the ice is left to rot by the warmth of the sun, unaccompanied by rains, it breaks up quietly, and with a slow current ; the force of the shove, even in this case, however, is enough to do very serious injury to temporary works. But if the breaking up is caused by rains, it is accompanied by a sudden rise of several feet, a rapid current, and destructive gorges; such was the case in 1867, when the ice broke up, on the 7th of February, with a sudden, though temporary, rise, and a current of six or seven miles an hour.

This behavior of the ice is not unlike that on all large rivers, while its movements are much more violent and destructive in colder climates than that of Kansas City; the only danger to be apprehended comes from the first five or ten miles of ice, as the rapid current and crooked channel soon break the larger fields into cakes too small to do much damage. The phenomena of the shifting channel and variable bottom were, on the other hand, in a large measure peculiar to the Missouri, and the difficulties which they must cause under the most favorable circumstances, were augmented by the imperfect understanding of them which existed; they had been made subjects of common report rather than of accurate observation; they were well known to old settlers and to river pilots, but the Government Surveys, to which we are indebted for our best information concerning the Misissippi and its other tributaries, had not been extended to the Missouri. The isolated character of the location, nearly three hundred miles from any general markets or machine shops, and a very much greater distance from the manufacturing centres of the country, was an additional source of trouble, involving the construction of an equipment which could elsewhere be bought or hired in a few hours, and causing occasional inconvenience by delays in the arrivals of material ordered from a great distance.

As Kansas City is situated immediately below the mouth of the Kaw, the Missouri is there affected by all the floods of the upper rivers, the Platte, and the Kaw, but free from those of the Osage and the lesser tributaries below. The distance between the bluffs is here about two miles, but the course of the channel is somewhat complicated by changes in their direction; the eastern bluff curves from its nearly north and south course to one bearing east and west; that which is the west bluff of the Missouri, a few miles above, becomes here the north bluff of the Kaw, leaving a considerable bottom land west of the city; east of the town the southern bluff follows a tolerably regular course. The channel has continued for an indefinite time at the foot of the bluff, in front of the city, but material changes have taken place above. The shore line of 1826, the oldest of which any record remains, differed 3,000 feet from the present shore line, the river having been steadily cutting into the Kaw bottom, above Kansas City.[*] The effect of this cutting has been to diminish the angle at which the current impinges on the bluff in front of the city, so that in 1867 the current line had become nearly parallel to the shore line at this point, though somewhat divergent to the north, causing it to wear away the opposite shore a mile farther down. A further continuance of this abrasion of the Kaw bottom, must have resulted in so sharp a bend as to throw the current against the northern shore at or near the bridge site, and this change of channel would have been accompanied by the deposit of a sand bar in front of the steamboat landing, and eventually, by the formation of a new bottom land between the city and the river.

A daily water record was kept during the whole progress of the work from January 1, 1867, to June 30, 1869.[†] The surface of the water at the former date was taken as a datum height and called 100; this served as the bench to which all levels subsequently taken upon the works were referred. This elevation may also be taken as the ordinary low-water level, though the extreme low-water mark is about three feet lower; a stage of 97.4 was observed on the 24th of December, 1867, and the low water of 1860 was probably a few inches below this. The height of the great flood of 1844, the highest flood of which any authentic record could be found, was pointed out in several places by old residents, the elevation of the water marks thus shown was carefully levelled, and the height of this flood referred to the datum line; the result of these observations proved that the water then reached the elevation 134.29, showing a range of 37 feet three and a half inches between the extremes of high and low water. The highest flood which has occurred since took place in 1858; but the water then rose no higher than 122, falling more than 12 feet below the flood of 1844.

The Missouri is subject to floods of greater or less magnitude during six months of the year, from February to July inclusive. The most violent of the early floods are to be attributed to the breaking up of the ice in the Kaw and the Platte, especially the latter; the others are due to local causes. In June

---

[*] See Map, Plate I.   [†] This record is given in profile on Plate II.

comes the mountain rise from the upper river, which usually attains its greatest height early in July; it is the most certain, and except on rare occasions the highest flood of the year. By the 1st of August the river has begun to subside, and it continues to fall, with few or no disturbances, till the low winter stage is reached. The best working season is from the middle of August to the middle of December, the winter work being made extra hazardous by the dangers from ice, and the spring freshets greatly curtailing the amount of work which can be done before the end of the summer.

Observations taken to determine the speed of the current showed a minimum velocity in the channel of two miles an hour—this being in the early part of February, 1868, when the river was frozen and the water extremely low. The greatest velocity accurately observed was in April, 1867, being at the rate of 12.7 feet per second, or a little more than eight miles and a half an hour. As the speed of the current is largely dependent on the rapidity with which a flood rises, and as the river has been known to rise at other times and places much more rapidly than was observed at Kansas City during the building of the bridge, it is likely that the maximum speed is considerably in excess of that given above, and it may be as great as 12 miles an hour.

The unstable condition of the river bottom was fully confirmed by soundings taken in the spring of 1867, at four successive times, from which cross-sections of the river were carefully plotted.* Changes continued to take place with equally marked effect during the two following years, but as their very frequency made each individual change unimportant no record was kept of the subsequent changes, beyond the soundings taken from time to time at the several pier sites and for special purposes.

In selecting a location for the bridge, it was necessary not only to place it where it should best fulfil its commercial requirements, and to see that it formed as slight an obstruction to navigation as was possible, but also to locate it at a point where a moderate amount of artificial shore protection would suffice to hold the channel permanently at the draw. As the bridge was designed to accommodate the high way travel as well as the railroad, it was necessary that it should be opposite the city, and a very favorable location four miles below

* See Plate II.

Engineered Irony

had to be rejected. The location adopted is a few hundred yards above the public steamboat landing, and crosses the river at a point where the channel, after sweeping round the long curve above the town, is still close to the Kansas City shore. The southern bank is here a rocky bluff, and rock was found in the channel near this bank only a few feet below the low-water level, with a local dip towards the north of about 1 in 20. The northern shore is the customary low Missouri bottom land, below the extreme high-water mark, though rarely overflowed. The width of the river, measured from the wooded shore on the north to the rocky bluff on the south, is almost exactly a quarter of a mile, thus making the crossing a conveniently short one ; at low-water the water way contracts to about 750 feet, leaving from 500 to 600 feet of sand bar between the water and the northern bank. When the location was made, the surface of this sand bar was at an elevation of 104 ; but it has been raised by the deposits of the three succeeding seasons, and is now 111. The southern portion of this bar, lying nearest the channel, is very variable, being liable to be washed out and replaced several times in a year ; but the northern portion, about 400 feet in width, seems to have become permanent, and unless disturbed by an extraordinary flood within a few years, it will become a part of the wooded shore. The channel lying along the south bank was regulated by the long curve in the river above, and would be rendered permanent by protecting the west or outside shore around this bend. This protection was demanded, not only by the bridge, but by the general interests of the city, the value of the bottom land being too great to allow it to be carelessly washed away ; while the lower part of the town, including the steamboat landing, was liable to be shut off from the river if the channel shifted any further to the north.

These merits of location may be briefly enumerated as follows :

First.—Proximity to business and the city.

Second.—Shortness of the bridge line.

Third.—Small depth to rock on south side.

Fourth.—Permanency of the channel, this being easily secured by protections demanded by other interests than those of the bridge.

The sole objection to this location lay in the southern approach, which made it necessary to leave the bridge on a sharp curve, and involved a cut 70 feet deep; these were matters of little weight when compared with the merits just enumerated.*

Great care was taken to ascertain the exact direction of the current and the bridge line. Observations were made with two transits simultaneously on a number of floats; the result of these observations was mapped out on a chart, and the direction of the current was thus accurately determined. The course in mid-channel as thus obtained made an angle of 72° with the bridge line; the piers were located parallel to the current, and the bridge built on a skew of 18°. The floats used were made by inserting a rod, to which a small flag was attached, in the neck of a bottle loaded with shot, the amount of shot being sufficient to sink the flag nearly to the surface of the water. These floats drew, on an average, about four feet, or the usual draught of a Missouri River steamboat; they therefore gave the true navigable current. The direction of the current varies a little with the stage of the water, as well as with the different forms which the smaller sand bars assume in successive years, but during the season of navigation it has not been found to show any material divergence from the line thus determined and given to the piers.

The pivot pier was placed in the centre of the channel; the piers were numbered from the southern end of the bridge, and the lengths of the several spans were as follows: a fixed span of 132 feet, extending from the shore to Pier No. 1; a pivot draw 363 feet long, each arm having a clear span of over 160 feet, as required by the Act of Congress; a fixed span of 250 feet from Pier No. 3 to Pier No. 4; leaving a remaining distance of 577 feet divided in the original plans into three spans of equal length, though subsequently changed to two spans of 200 feet each and one of 177 feet; to this must be added a shore span of 68 feet at the south end, extending over the width of a street and the Pacific Railroad track, and which made the total length of bridge from outside to outside of masonry 1,400 feet. The lengths of spans given here are gross distances taken from centre to centre of the adjoining piers. The nearness of the channel to the south bank made it impossible to place a span of 250 feet

---

* The profile and alignment are given on Plate II.

Engineered Irony

on each side of the draw, as required by the Act of Congress, without neglecting the more important provisions of that Act, which requires the draw to be placed over the centre of the navigable channel. The requirement of two spans of that length was made, as has already been stated, with special reference to the Mississippi, where raft navigation forms an important part of the river commerce; on the Missouri, raft navigation, except on the most diminutive scale, is impossible, owing to the sharp curves, strong current, and multitude of snags, while no sufficient supply of good timber is found along the river to make rafting profitable, even if it were possible. Endeavoring to conform as nearly as possible to the Act, a span of 250 feet was placed immediately north of the draw, but the unexpected destruction of the foundation works of Pier No. 4, in the spring of 1868, made a further change necessary; the site of the pier was moved 50 feet farther south, and the long span finally built between Piers 4 and 5.

The month of April, 1867, was distinguished by a very extraordinary spring flood, caused by the united freshets of the Platte and Kaw; on the 29th of the month the river had risen to 119.3, three feet and three inches higher than the June flood of that year, and four feet above the highest water of either of the two succeeding years. During the war the Government had established a supply station in the bottom west of the town, and a small portion of the bank had been protected by a covering of riprap, to serve as a steamboat landing. The river had washed the bank away both above and below this protection, leaving a projecting point, the end of which alone was covered with stone; though the whole discharge of the river passed outside of this point, strong eddies were formed on each side of the neck of land connecting it with the shore, which gradually reduced it to a long and narrow isthmus. On the 10th of April the width had been reduced in places to only 18 inches; on the following day the two eddies met, entirely destroying the neck of land, and allowing the channel of the river to shift at once from the north to the south side of the rocky point, changing its position nearly 500 feet in a single day. The pile of rocks still remains in its old place, having been left almost dry on the north side of the river during the low water of December, 1867. As the channel shifted suddenly from the north to the south side of it, it has never been exposed to the most violent force of the current, but its height has gradually diminished, and it will in time sink out of sight.

The stone protection above the bridge was begun during this flood, and continued through the summer and early autumn of the same year ; it was carried westward from the point where the rocky bluff and shore line separate above the bridge, to within 150 yards of the State line. This protection was executed under the direction of the engineers, a portion of the expense being borne by the city. It consists of a simple revetement of riprap stone, a large portion of the stone used having been taken from a cut in the southern approach to the bridge. Whenever practicable, the shore was first worked by laborers to a slope of about one to one, and the stone was evenly distributed over this slope ; but when the revetement was begun, the water was too high to admit of this, and the stone was simply dumped over the bank, till the heap appeared above the surface of the water. This protection has required some repairing, the stones having slid down the slope, replacing the soil which the river had washed from beneath them ; but it has proved perfectly effective, and the river has in no instance changed the line of the protected shore. In the following spring the protection was extended to the State line, and during the low-water season of 1868–69, it was carried, in the interest of the land owners, as far as the mouth of the Kaw.

The railroad approaches the bridge from the north, with an ascending grade of one in one hundred, till within 618 feet of the bridge, after which the track is level ; this leaves room for a train to stand between the grade and the bridge. The 2,380 feet of this approach, adjoining the bridge, is an open trestle work, thus making an effective water-way of 3,775 feet in times of extreme flood, when the bottom land is overflowed. The trestle is substantially built of native oak timber ; the 50 bents nearest to the bridge rest on piles, and the others, 90 in number, are on sub-sills. The roadway approach is by a side trestle, built out on the west from the bents of the railroad trestle ; it has a grade of four in one hundred. The two approaches unite at the second bent from the bridge, where is placed a toll-house and gate .* This trestle was built by the

* See Plate IX.

Company at such intervals as the carpenters could be most advantageously employed upon it, in 1867 and 1868.

The carriage road was continued south of the bridge only far enough to allow teams to turn off into the adjoining streets. The railroad approach leaves the bridge on a 9° curve to the right, this curve beginning at the middle of the 132-foot span ; with a maximum descending grade of 42.24 feet in a mile, it passes through the bluff in a clay cut 72 feet deep, and then passes down along a rock cut in the west side of the bluff to the depot grounds in the West Kansas bottom. The grading of this approach was let by contract and the work completed during the year 1867.

CHAPTER III.

FOUNDATIONS.

FROM the inception of the work the subject of foundations was the paramount study of the engineers, the only real difficulties of the task lying below the water. The methods of founding which have been in most common use in the United States were not to be thought of, as the continual wash and scour of the river would have made piles and crib-work useless, while the great depth and rapid current must have rendered coffer-dams very hazardous and expensive. The use of iron columns, sunk by the pneumatic process, was considered; but the conviction was early and confidently formed that a cluster of separate columns resting upon the rock at a depth but little below the scour limit, as would have been the case in the most exposed foundations at this location, would fail to give the stability needed by the channel piers; while it was believed that the sand-bar piers, which are rarely exposed to a strong current, might be founded in a way less expensive, though amply secure. It was also feared that in the absence of pneumatic plant in America, and with the then high prices of iron work, the pneumatic process would prove in its entire execution an unreasonably expensive one.

The opposite action of floods on the two sides of the river, causing a violent scour along the Kansas City shore where the channel lies, but a sand deposit near the north bank, showed that the precautions necessary for the channel piers would be a useless expense if taken at every foundation. The channel must be retained near the south bank on account of the draw, if for no other reason—thus rendering this phenomenon of scour and deposit a permanent one, subject only to such variations as are due to the increased width of channel in an extreme flood. It was therefore thought that if the channel

5

piers were founded directly upon the rock, the others might safely be put upon piles, the care taken to protect the pile foundations increasing with their nearness to the channel. This arrangement was believed to have the farther advantage of making it practicable to begin work on the sand-bar piers at an earlier date than would otherwise have been possible. It was accordingly determined that Piers Nos. 1, 2, and 3, should rest directly upon the rock, while the four more northerly piers should have pile foundations; the piles were to be driven in excavated pits, cut off in every instance at a considerable depth below the usual bed of the river, and further secured by an ample protection of riprap; those under Pier No. 4 were to be driven home to the rock, and cut off at least 25 feet below extreme low-water. The experience acquired during the progress of the works led to subsequent changes in the plan; Pier No. 4 was treated as a channel pier, and founded on the bed rock, and the piles under Pier No. 5 were driven home to the rock, thus reducing the number of pile foundations to three, only two of which depend upon the frictional surface of the piles for their support. Piers Nos. 5 and 6 are on the dry land of the sand bar for nine months in every year, and Pier No. 7, situated within the line of the wooded shore, is exposed to the action of the water only on rare occasions. During the progress of the works no current strong enough to produce scour was noticed about any one of these piers.

All the foundations were put in of a sufficient length for a double-track bridge.

Although the foundations may properly be grouped in the two classes of channel and sand-bar foundations, as above mentioned, the characteristics of several pier sites were so different that it became necessary to treat each by itself, and to prepare as many sets of plans as there were piers. The south abutments and the two small pillars on the bank were built on the rock which was found a few feet below the surface of the ground, their foundations presenting no greater difficulties than are common to every cellar wall. They were built in the fall and early winter of 1867–68, the work upon them being executed at such intervals as the masons were not employed upon the river piers. The other foundations were taken up in the order of convenience; the dates at which the work in the river was begun in each instance, and the dates at which the several foundations were ready for the masonry, were as follows :

Pier No 1, Water Deadener sunk ....Aug. 16, 1867.	Masonry begun ..Oct. 16, 1867.
" 2, Caisson launched.............Sept. 23, 1868.	" " ..Feb. 20, 1869.
" 3, First Pile driven in False Works Aug. 27, 1867.	" " ..May 20, 1868.
" 4, " " " " Sept. 2, 1867.	" " ..April 2, 1869.
" 5, Erection of Caisson begun ......Jan. 20, 1868.	" " ..Jan. 14, 1869.
" 6, First Sheet Pile driven........Sept. 3, 1867.	" " ..Jan. 3, 1868.
" 7, First Pile driven......... ......Feb. 27, 1867.	" " ..Oct. 1, 1867.

These dates do not in themselves give a correct idea of the time actually consumed in putting in the substructure of the piers ; the work on some of them was more than once suspended, either on account of the season or to accommodate other work which had momentarily become of greater importance. After the caisson had been built at the site of Pier No. 5, it was left standing for two months, until machinery could be spared to sink it. At Pier No. 4, the first set of works was entirely destroyed in the spring of 1868, in consequence of which the site of the pier had to be changed, a new plan of foundation prepared, and the first pile in the new false works was not driven till the 9th day of August, 1868 ; moreover, the date given above as that at which the masonry of this pier was begun, is in reality that at which the laying of masonry was resumed after the completion of the foundation and the removal of the upper false works ; the bulk of the subaqueous masonry here was put in during the process of founding. In the cases of Piers 1 and 2, the caissons were built on shore, and the time occupied in the foundations, including this preparation, was therefore greater than the table indicates.

A correct idea of the time occupied in the different foundations, as well as of the characteristic features of the several plans, can only be given by separate narratives of the substructure operations at each of the seven pier sites.

PIER No. 1.

This pier is situated on the south side of the main steamboat channel, and at ordinary stages of water it stands about 100 feet from the shore line. The bed rock was found at an elevation of 84, or 13 feet below the extreme low-water mark ; at the pier site it was of irregular form, and though found to fall off rapidly a few feet farther north, it here presented a surface that was almost level, though inconveniently rough. This rock was almost bare, being seldom covered with more than a foot or two of deposit. The current was but little less rapid than in the middle of the channel, and was far too strong to allow of any extensive operations being carried on within it. It was therefore thought necessary first of all to obtain slack water about the pier site, after which the foundation works, more properly so called, could safely proceed.

A large timber caisson, designed to serve as a water deadener, or break-water, was built on the shore, about midway between the south end of the bridge line and the Company's machine shop. It was built of oak, with pointed ends ; the entire floor and the sides to a height of 15 feet were solid, and of square timber ; its outside measurements were 65 feet long from nose to nose, 18 feet wide, and 27 feet and three inches high ; it was stiffened internally by rows of vertical truss bracing, and bound together by long iron rods built into the solid timber of the bottom and sides ; the whole was thoroughly caulked, and valves were provided for admitting or excluding the water.* After it had served its purpose as a water deadener, it was raised and finally sunk below Pier No. 2, where it forms the foundation of the lower draw rest. On the 19th of May, 1869, this caisson was launched ; it was kept anchored to the shore till the 7th of the following August, when the river was thought to be low enough to begin work ; it was then towed to a point about 100 feet above the pier site, secured by four wire cables reaching to the shore, and sunk by admitting water through the valves and throwing in a ballast of broken stone. Being placed transversely with the current, it formed a complete water deadener, and quiet slack water was secured at the pier site.

A bottomless caisson, which should serve as an enclosure to build the pier in, was also constructed. It was built on a floor placed between four boats well braced together ; it was a frame structure, entirely of oak and planked vertically ; the ground plan was substantially the same as that of all other caissons built upon the work, the ends being formed of two short sides making a right angle together ; its total length was 70 feet, and the width 19.5 feet. The first

* The plans of this caisson are given on Plate III.

section was 13 feet high, and as soon as this was completed it was lowered between the boats almost to the surface of the water, and a second section, 11 feet high, added to it. The lower edge was shaped to fit the irregularities of the rock, and the caisson was surrounded by a sheet piling of planks sharpened to a feather edge and secured by a double set of guides. The caisson was thoroughly caulked, braced internally, and fitted with a false bottom; this bottom, which was put in to facilitate handling, was built in sections, placed a little above the lower edge, and held in position by inclined braces bearing against the caisson timbers and adjusted with folding wedges. To aid in placing the caisson upon the rock, eight posts were provided, each 60 feet long and 16 inches square, built of 8 by 16 oak timber of shorter lengths, with a central hole three inches in diameter extending from end to end, through which a two-inch steel drill, welded to an iron rod 65 feet long, might be worked.

When complete, the caisson, still carried by the boats, was carefully floated into position, and four of the 60-foot posts with the drills, raised on either side of it. The posts sank by their own weight through the thin layer of sand, and were at once made fast by working the drills two feet into the rock; they were then well braced together and secured to the caisson by sets of rollers and shoes, the two sets of posts being placed about an inch nearer together at the top than bottom, to secure clearance in lowering the caisson. Four pairs of cross timbers, attached to the posts, were placed above the caisson, each of them carrying two suspension screws 20 feet long and two inches in diameter, with a thread cut from end to end. The total weight of the caisson being 72 tons, each screw was required to carry nine tons. On the 6th of September every thing was in readiness, the caisson had been attached to the screws and the lowering was begun. Three men at each screw were required to handle the weight. Ease and regularity of descent were secured by admitting water above the false bottom; when one half of the caisson had become submerged this was found to be no longer necessary, and the bottom was set free by striking the folding wedges which held the braces, and taken out in parts. On the 11th of the same month the caisson came to a bearing upon the sand, and the screws were removed. The use of long posts secured by drills rendered the matter of false works exceedingly simple; this device, which is believed to have been entirely novel, is admirably adapted to use in places where piles cannot be driven, and posts merely braced together are not to be trusted ; it is effective, can be put in rapidly, and may be used where the rock is overlaid with several feet of sand, the posts being sunk by blowing away the sand with a stream of water forced through the central hole.

After the removal of the screws the caisson was sunk two feet farther to the rock by means of a water jet, directed by a diver working on the outside ; a single day proving sufficient to bring it to its permanent bearing. The water jet used on this, as well as on subsequent occasions, consisted of a copper or iron nozzle attached to a three-inch flexible hose, and sufficiently loaded to be handled with ease under water ; through this a stream of water was driven by either a centrifugal or reciprocating pump, the former being used on the foundations first put in, but the latter being found the more efficacious. The attempt to fit the rock by shaping the bottom of the caisson proved a failure, and a good joint was only secured by means of the sheet piles ; these were driven by a light ringing engine, and by bruising against the rock secured an accurate fit. The caisson was then surrounded on the outside by a double row of gunny bags filled with clay, which were carefully placed by the diver in a trench excavated by him with the water jet. These bags were further surrounded with hay, which was again covered with a bank of clay, protected from the water by a canvas tarpaulin, and the whole covered with a layer of clay and stones.

On the 10th of October these preparations against leakage had been completed, and the work of pumping out the caisson was begun. The joints were found to be admirably tight, a nine-inch Alden pump, driven by a twelve-horse power engine lowering the water five feet in an hour. Additional braces were placed within the caisson as the water was lowered, to resist the increasing outside pressure. The small amount of sand and mud remaining on the rock was cleaned off, and removed in boxes, when two beautiful springs of clear water, contrasting strongly with the muddy Missouri, were found issuing from the fissures in the rocks, and the rough and jagged surface was quarried to an even bearing, suitable to receive the masonry. The solid character of the rock was also proved by drilling into it.

To facilitate handling the stone a trestle bridge was built, extending from the shore to the south-west corner of the caisson, and on this were laid two tracks of wooden railway. A floating derrick, the larger of the two used, was anchored on the south side of the caisson, just below the bridge ; two small cars pushed by hand brought the stones within the range of this derrick, by which they were placed at the desired spot within the caisson. The first stone was set on the 16th day of October ; a belt of masonry was first laid around the caisson wall ; this was then backed up and the masonry above built up in regular courses ; the space between the face of the stone and the planking of the caisson was filled with beton. As a precaution against the accidents which needless delay might involve, the work, during the first week, was driven both night and day. The leaks were soon found to be so slight that the use of the steam pump was unnecessary, and after the first few courses had been laid the water was kept down by two common log pumps worked by hand. The work upon this and subsequent piers (with the exception of No. 4) were directed by Mr. W. K. McComas, superintendent of foundations.

PIER No. 2.

With a view to avoiding any delays which might arise from unforeseen difficulties attending work in the main channel, where the dangers of accident were thought to be greatest, as well as to secure the greatest possible time for raising the draw, the preparations for the building of this pier were among the first taken in hand. The principal difficulties lay in securing staging to work from at the pier site ; piles could not have been used, as the rock was frequently swept almost bare of sand ; anchored posts, such as had been used at Pier No. 1, might have been destroyed at any moment by an accidental blow from a descending steamboat ; while long and wide cribs, such as have been used on the Mississippi, would have been unmanageable, and have taken up too much water-way in the narrow channel of the Missouri. Small detached cribs and caissons were finally used, it being hoped that the current would remove most of the sand from beneath them, if they were held floating in position while a scour took place below ; much difficulty was experienced even in handling these small bodies, and the scour took place so irregularly that some of the intermediate cribs were finally lost. Early in the spring of 1867, work was begun upon a timber crib, which was to form the permanent foundation of the upper draw rest, and serve as an anchorage and water deadener while putting in the foundation of the pier. This crib was built upon the sand bar on the north side of the river, and launched by the rising water of the April flood; the building was continued after it was fairly afloat, and it was poled out into deeper water, from time to time, to prevent grounding. One Sunday morning, while this work was still in progress, a steamboat going up the river, in hugging the shore to avoid the strong current of the channel, fouled against one of the lines, which breaking released the crib. The yawl crew, who were watching for any such accidents, at once boarded it, and, being unable to make fast to anything, went down the river with it. As soon as the crew of the steamboat could be collected and steam raised, she was sent in pursuit, but having lost an hour in the start she did not overtake the crib till it had drifted 22 miles down stream. Two days having been spent in a fruitless attempt to tow the crib against the current, it was taken apart and the timbers brought up on barges.

The caisson in which the pier was built, was built in the summer of the same year on the south bank of the river, in front of the Company's machine shop. Its form was that of a round tub 18 feet high and 40 feet in diameter at the base, the sides sloping inwards with a batter of 1 in 16. It was made of four-inch oak staves, six inches wide, bound with flat iron hoops, and strengthened by timber rings on the inside.* The bed rock at the site of this pier is found at an average elevation of 80, with a slope towards the north of nearly three feet in the diameter of the tub. When the first soundings were taken this rock was found overlaid with eight feet of sand, to secure an easy penetration through which, the caisson was provided with an iron cutting edge, formed of pieces of three-eighth inch boiler plate riveted together and fastened to cast-iron brackets, which were bolted to the lower set of internal rings. These plates were trimmed off so as to make a difference of 18 inches in the heights of the opposite sides of the caisson, thus partially balancing the slope of the rock. The tub was provided with a false bottom, built in radial sections, suspended at the centre from a light truss overhead, secured by inclined

---

* This caisson is shown on Plate III.

braces and folding wedges. Under it were placed three shoes, which slid on inclined launching ways, provided with guides for their whole length.

In October, the crib for the upper draw rest was rebuilt on the south bank of the river, a third of a mile above the bridge ; it measured 31 feet wide, 73 feet long, and 10½ feet high ; was built of square timber, with cross walls and bottom of round timber and planked on the outside ; the upper end was formed like that of the piers and caissons, and the lower end square, though shelving forwards—this form being thought favorable to a scour beneath ;* additional buoyancy was secured by binding a number of empty coal oil barrels inside of it. Early in November an attempt was made to launch the crib, but its misfortunes were not yet over, the ways breaking down under the weight and leaving it lying on one side at the edge of the water ; nothing could be done towards raising it in consequence of the rapidity of the current in front of it ; and in this position it remained till the following February. On the 9th of January the ice jammed at the bridge site, and the river for several miles became closed ; a channel was cut in the ice from the bank where the crib lay to its permanent location above the pivot pier ; on the 4th of February it was raised by means of hand crabs, and successfully floated into position ; the water was very low, the current only two miles an hour, and the crib was easily held by lines attached to posts on the shore and to anchors put in the ice. It was sunk by putting on additional courses of timber, and throwing in rubble stone ; the current swept the sand, about 10 feet in depth, away from below, and allowed it to settle firmly upon the rock. As this crib had to be placed in position before the other works connected with the pivot pier could proceed, the difficulties in the way of locating it accurately, with no neighboring anchorage, were very great, and the misfortunes of the launch in October may have been more than compensated by the advantages which the ice gave for handling the crib when it was finally placed. On the 17th of February the ice gave way above the bridge and went out, doing no further damage to the crib than to loosen the upper course of timbers and fill it with ice. Before the close of the low-water season the more exposed parts of the crib were filled with beton, and additional stone was thrown into the central divisions.

---

* The shape of this crib appears on the plan of Draw Protection, Plate VII.

Engineered Irony

The delays due to this series of accidents had consumed almost the whole low-water season, and it was thought unwise to proceed with so exposed a foundation till after the summer flood. In the following September the water-deadener caisson was raised from its place above Pier No. 1, and dropped below the site of Pier No. 2, where it forms the foundation of the lower draw rest. The floods of the preceding summer had filled it with several feet of mud, which had to be removed, when it was easily raised by pumping out the water. No small difficulty was experienced in getting it accurately placed, owing to the inconvenient distance of the anchorages, and frequent interruptions from passing steamboats; it was finally secured within a few inches of the desired spot, though not quite parallel with the line of the piers, an irregularity which was taken out in the framing of the upper works. An additional amount of stone was thrown in and an unyielding bearing obtained on the bed rock, which was swept clear of sand by the current.

On the 23d of September, 1868, the round caisson, which had been ready on the launching ways for nearly a year, was successfully launched. The bottom and caulking proved tight, and no pumping was required to keep it afloat. An additional section nine feet high was put on, built of oak staves with hoops and rings, and differing from the lower section only in having no batten bottom;* a wall of rubble masonry was laid between the rings of the lower section to give weight. The caisson was then dropped around Pier No. 1, and swung into position; it was handled entirely by lines, made fast to the two draw rests and adjoining piers, no false works being erected at the pier site. It was sunk by admitting water above the false bottom, and in two hours settled to a bearing. By pumping out, and readmitting a portion of the water, it was several times raised and lowered a few feet, until, on the 11th of October, it was brought into its final position, and firmly grounded on the rock. Soundings taken a few days previously had shown a deposit of from two to five feet of sand; but this was soon swept away by the current, when the iron edge was raised a few inches above its surface. Besides the hempen cables used to handle the floating tub, a two-inch wire cable was made fast to the upper draw rest, and attached by a bridle to the tub, being made two feet and

* This section is shown with the first, on Plate III.

a half short, to allow for any rendering under a strain; the strength of the current, which was but slightly deadened by the upper rest, was so great that this allowance proved insufficient, and when finally placed, though almost exactly in position on the bridge line, the caisson was found to be about 20 inches too far down stream—an error which, being less than the excess of the radius of the caisson over that of the pier, was of no consequence.

The caisson was now surrounded with a row of gunny bags filled with clay and packed around with hay ; the current was too strong to allow a diver to work on the outside except on rare occasions, and the bags were placed by lowering a number of them, united into a string, from above ; a joint formed in this way must, however, be very imperfect, and it was but little relied on. An inspection of the cutting edge, made by a diver within the tub, showed that, in consequence of the irregularities of the surface, it was in actual contact with the rock at but few points, while in some places the distance between was as much as 10 inches ; it was also found that a cone of sand was left inside of the caisson three or four feet deep at the centre, and which was constantly increasing from the deposits swept under the iron edge. Holes drilled four or five feet into the rock showed it to be perfectly substantial. A circle of gunny bags filled with freshly mixed beton was then placed by the diver against and under the inside edge, over which was spread a ring of beton of triangular section ; the beton in this ring was lowered in boxes from above and placed by the diver, who first carefully jetted away the sand remaining on the rock.

On the 29th of December, after waiting two weeks for the beton to harden, an attempt was made to pump out the caisson. The joints were found to be perfectly tight, the subsidence of the water being nearly equal to the discharge of the pumps; but when the water had been lowered nine feet, the outside pressure broke through a fissure in the rock under the beton and forced a leak which exceeded the capacity of the pump. An additional amount of beton was put in, increasing the section of the ring, and the crevice in the rock filled up as well as possible, and after a short delay a second attempt at pumping was made, but with no better results than the first. Not wishing to lose any more time, no further attempts were made to lay bare the rock; after pumping out the sand in the middle of the tub, a beton foundation eight feet deep was put in,

Engineered Irony

covering the whole circular area. This was in accordance with the original intentions of the chief engineer, the plan of starting the masonry on the bed rock having been an after-thought, and regarded throughout as an experiment. If precautions of the same kind as at Pier No. 1 could have been taken, this plan would probably have been successful; but it was first determined to make the attempt after the caisson had been placed, when it was too late to arrange the sheet piling which had proved so effective before. In the haste, also, to have this foundation completed, an insufficient time had been given for the beton to harden, showing the danger of haste in work of this kind.

The bulk of the beton in this foundation was laid in a box designed expressly for these works.* It was of rectangular form, having a capacity of half a yard, with a bottom formed of two leaves opening outward from the centre and fitted with a cover of similar construction. On either side two chains, one fastened to each of the lower leaves, were united into one near the top of the box, and those from the two sides were brought together in a ring hung upon a tripping-hook, of no novel form, above; the tripping-hook was permanently attached to a cross-head by a loose chain. When the box had been filled with beton the covers were closed, and it was lifted by a rope attached to the short end of the tripping-hook and lowered into place; the tripping-hook was then raised by a hand-line attached to the long end till the ring dropped from it; on drawing up the main rope the box was lifted by the loose chain attached to the cross-head, the leaves opening freely and discharging the beton. The advantages of this form of box lie in the protection it affords from washing in the descent and discharge, and the ease and certainty with which it may be tripped when once lowered upon a bottom.

On the 24th of January the pump was started and the water in the caisson was lowered rapidly. The following day the tub was pumped out and the beton laid bare; it was found in general to have set satisfactorily, though not perfectly homogeneous, and covered with several inches of *laitance;* but after a few hours the water came in again, probably working its way through fissures in the rock and voids in the foundation. Grout was poured through a funnel into the holes through which the water had come, as far as this was found possible, and six

---

* This box is shown on Plate VI.

additional feet of beton, making in all 14 feet, put in. The outside of the caisson was examined by a diver, who reported that the bags had been swept away from the south side, leaving an aperture under the edge, which was closed with bags of beton. About the middle of February the tub was again pumped out ; the leaks were still very troublesome, but within the capacity of the pumps ; the surface of the beton was levelled off, and an open grillage, composed of two courses of flatted timbers laid transversely, put in ; the spaces between the timbers were filled with beton, and on this grillage the masonry was started on the 20th of February. The rising water threatened to drown out the works, and a third section nine feet high was added to the tub, the river rising above its base for one or two days. This section was similar to the one below, but built of pine ; it was afterwards removed and made into a railroad water tank. The base of the masonry is at an elevation of 95.57 ; the first course of stone is the full size of the tub, and from this the courses are stepped off till their diameter is reduced to that of the pier. The floating derrick used to lay the masonry was fastened on the north side of the caisson, and the pier built up as rapidly as possible.

PIER No. 3.

Work was begun on this foundation on the 29th day of August, 1867. The rock was found at an elevation of 67—30 feet below the extreme low-water mark—and was then overlaid with 22 feet of sand, the water being 17 feet deep. A compact cluster of piles was first driven, 150 feet above the site of the pier, to serve as an anchorage during the subsequent work ; one or two of this clump washed out before the driving was completed, but the rest were secured by immediate riprapping, and have remained firm for more than two years. The pile-driving boat was then dropped below, hanging to the anchor piles, and twelve piles driven to form an instrument stand for use in locating the pier. But it was found very difficult to make a pile stand at all in this rapid current ; the rush of the water swayed the head of the pile back and forth several feet, washed around its base, and dug out the surrounding sand till the pile popped up and floated away. One pile in four of those driven was lost in this way with almost complete regularity. The piles were accordingly secured with some

Engineered Irony

difficulty, by bracing them together with planks; they were then cut off and a platform was built on top of them. The platform was found to sway five or six inches with the current, a motion which was reduced as much as possible by additional bracing; but on placing a transit upon the platform it was still found to vibrate more than an inch, and therefore to be wholly unfit for its intended purpose. The pier was accordingly located by measuring with a steel wire from Pier No. 1.

The pile driver was then moved to the pier site, and the piles driven for the false works proper.* The first of these piles washed out almost as soon as driven. It was evident that piles could be held here only by immediately bracing them together; for this purpose a plank was bolted on the side of the pile by a single round bolt of inch iron, at such a height as to be near the surface of the sand when driven home; the plank, being left free to turn on the bolt, was kept upright by lashing it to the side of the pile; after driving, it was to be swung over and spiked to the top of the next pile below. Several piles were tried in this way, but they all broke at the bolt hole under the concussion of the hammer. An arrangement was then adopted which had been successfully used at the bridge over the Rhine, at Coblentz. An iron ring, to lugs on the side of which a long iron rod was fastened by a pin joint, was dropped over the head of the pile when driven, and the rod made fast to the top of the pile below by means of a stirrup;† when found inconvenient to slip the ring over the top, it was made of two parts, which were bolted together around the pile; in spite of this precaution one-fourth of the piles driven were lost. The distance between the inside rows of piles was made 10 feet greater than the proposed width of caisson, thus leaving five feet for clearance on each side—an allowance which proved insufficient, as the piles, disturbed by the current and bruised against the rock in the driving, were sometimes forced considerably out of place, and thereby interfered with the caisson in its descent. The difficulties which attended this work at this favorable season showed that it would have been impossible if attempted during the floods of the previous months. .

When the piles had been driven and secured in this manner they were cut off and capped, and a floor was placed over the whole. Upon this floor the erection of the caisson was begun on the 20th of October; it was made similar in

---

* For plan of these works see Plate III.   † See side elevation on Plate III.

plan and shape to that used at Pier No. 1 ; but, being intended to penetrate through a considerable depth of sand, it was provided with a boiler plate cutting edge of the same kind as that used on the round tub, and the sides were given a batter of 1 in 16 ; it was thoroughly caulked and furnished with a false bottom. Four timber trusses resting upon the piles were placed above the caisson, transversely with the stream, which carried the eight long screws already used at Pier No. 1.* On the 12th of November the first section of the caisson was lifted from the floor by the screws, the timbers under it were removed, and it was lowered within a few inches of the water ; a second section was at once built upon it. On the 25th of the same month the caisson was lowered into the water till almost in contact with the sand, being held against the current by a wire cable attached to the anchor piles above. The false bottom helped, as at Pier No. 1, to secure ease and uniformity in the descent ; it was also expected to increase the scour immediately below.

The current rapidly washed out the sand to the depth of 15 feet at the upper end of the caisson, but only disturbed it slightly at the lower end ; the upper end was therefore kept hanging on the screws, and the lower end let down upon the sand. The false bottom was struck at once, and another section built on the caisson ; this third section differed from the two below in being planked horizontally and having the sides plumb. The lower end was then sunk about two feet into the sand by using a water jet around the edge. On the 10th of December a six-inch siphon pump, which had previously been in use at Pier No. 4, was placed in the caisson and worked by the boilers of the steamboat ; it threw enough sand to sink the caisson about a foot in a day at the first, but this rate of descent slackened to about six inches and even less, when the upper end reached the sand and the edge took a bearing all around. On the 30th of the same month a small dredge and a four-inch Andrews centrifugal force-pump were added to the outfit ; the dredge was of the endless chain pattern, mounted on an incline, and worked by four men with two cranks ; it had a capacity of about 50 cubic yards a day. The Andrews pump was placed inside the caisson and driven by a steam-engine on a boat swinging below the works ; though used to some small extent as a sand-pump, it was chiefly relied upon to drive a

* Shown on Plate III.

48   THE KANSAS CITY BRIDGE.

water jet, by which the sand from the more remote parts of the caisson was fed
to the siphon and dredge. To give additional weight, the spaces between the
timbers of the upper section were filled with rubble masonry, which was protected from injury by an inside sheathing of thin boards. In order to diminish
the external sand friction, a large cast-iron pipe was placed around the caisson
above water, from which a series of small gas pipes extended down nearly to the
cutting edge, thus forming a line of water jets encircling the caisson, and distri-
buting the discharge of a single pump around the whole surface. A nine-inch
Alden pump was attached to this set of pipes, but it proved unequal to the task,
and only half of the pipes, those on one side of the caisson, were ever worked
together ; the force of the stream so widely distributed was also too much
reduced for effective work, and it was generally preferred to use a single
movable jet, which could be taken in turn to every part of the caisson.

The changes in the bottom during the work on this foundation were very
frequent and singular, sometimes causing no small trouble. In November a
scour was noticed around the anchor piles, which was accompanied by a deposit
at the pier site ; the elevation of the sand, which was materially lower on the
first of that month than it had been when pile-driving was begun in August, was
raised 11 feet in three weeks. On the 2d of December there were 14 feet of
water outside of the caisson at the upper end, and 22 feet at the lower end ; one
week later this was reversed, when the water was found to be 24 feet deep at
the upper end and only 12 at the lower. Some of the smaller changes, though
less embarrassing, were more remarkable. About the middle of December the
river gouged out a hole in the sand, close to the south-west corner of the caisson,
extending clear to the rock, which was examined by the diver ; the sand was
soon filled up to the ordinary level. A similar hole was washed out about a
week later, and some pieces of floating ice were sucked under the edge of the
caisson and came up inside.

On the 10th of January the descent was stopped by coming in contact with
a buried log ; this was cut half through by the diver with a handsaw, and then,
the saw binding so much as to make its farther use impracticable, a large chisel
was mounted with a long handle so as to be worked from above, and the cut
finished by striking repeated blows on it ; only four days were lost by this ob-

struction.   A fortnight later the caisson struck upon one of the bearing piles, which was pulled up from the outside, the platform over it having first been sustained by trussing across between the two adjoining piles.   At this depth the work was also occasionally impeded by sand slides, the first notice of which was given by a sudden rise of the water inside of the caisson, a considerable quantity of sand from the exterior slipping through under the cutting edge and forcing the water before it, which increased the amount of excavation required and engendered a corresponding delay.

On the 6th of January very cold weather had set in ; three days later the ice had jammed at the bridge line, and the river closed for the season.   The ice soon became so strong that the stones used in the masonry at Pier No. 6 were taken across on wagons.   To guard against the injury which might result from the breaking up of this ice, the cluster of anchor piles was made into an ice-breaker by surmounting it with a small triangular crib of square timber, to which was fastened the upper end of a single inclined stick, the lower end of which rested on the bottom of the river 15 yards above, being held down by another small crib filled with stone ; a row of spring piles was also driven, extending from this ice-breaker to the instrument stand.   A narrow opening, about two feet wide, was cut in the ice on the channel side of these protections, extending some distance above them, by which the main field of ice was separated from that nearer the shore which was attached to the piles.   It was hoped that the fields of ice would be broken into small cakes by the inclined timber, and that these cakes would jam against the spring piles, pack to the bottom of the river, and thus form a gorge above the pier site which should protect the caisson from the shocks of the remaining ice.   The trusses above the platform were taken down, and everything not absolutely needed was removed to a place of safety.   About the same time a fourth section was added to the caisson ; it was in all respects similar to the third section, the walls filled with rubble masonry, and it made the total height of caisson about forty feet.

On the 12th of February, the upper end of the cutting edge came to a bearing on a point in the rock.   The day previous the weather had become warm, and the ice began to rot ; on the 17th, at about noon, it broke and moved down the river.   The water was very low, 100.5, the current less than three miles an hour and much of the ice quite soft, but it moved in large fields, and but for the preparations made, would have done great damage; it broke off and carried away nearly all the spring piles, completely demolished the instrument stand, and tore out some of the false-work piles on the south side of the pier; but the gorge hoped for was formed, and the caisson left uninjured.

Dredging and pumping were resumed, the weight upon the caisson was increased by piling heavy stones on the top, and a small pile driver was set up upon it, with which short blows were struck, with a view to loosening the sand pressure by the jar. By the 25th of February the whole upper half of the cutting edge rested on the rock, and six days later the desired bearing was reached around the entire caisson. A week was spent in removing the sand inside, when a tight joint was made by placing bags of freshly mixed beton against the interior edge in the same manner as was afterwards done at Pier No. 2. The rock was examined by drilling in it, and found to be solid and firm. There still remained a small quantity of loose sand, which was only removed by constant pumping, with the frequent attendance of a diver, keeping the water level constant by admitting water above.

On the 20th of March the rock was sufficiently clean, and the work of laying the beton foundation was begun. The beton was lowered in triangular boxes, similar to those used on the Quincy Railroad bridge, from which works the pattern of box was taken. The work was suspended on the 20th of April, fifteen feet of beton having been laid at that date. One week later the pumps were put in, and the water lowered; but as the caisson showed signs of yielding above the beton under the water pressure, pumping was stopped, and seven feet more of beton put in, making the full depth of the foundation twenty-two feet. This was concluded on the 5th of May, and on the 12th the caisson was pumped out; the beton used in this pier contained a greater proportion of sand than was used elsewhere, and was found not to have become entirely hard; the surface was therefore cleared off, and an open grillage built upon it, the spaces in which were filled with beton, and which was secured by iron straps to the sides of the caisson. On this the masonry was started at an elevation of 90.4, the first stone being laid on the 20th of May. The pier was at first built up only to an elevation of 108, it being thought best to give the beton an additional time to harden before exposing it to the full weight, while the summer flood, now close at hand, made it desirable to postpone any further work at this site. On the 10th of August the work was resumed, and the pier rapidly built up to completion.

PIER No. 4.

Work was begun upon this foundation on the 2d of September, 1867, four days later than at Pier No. 3. The water was then twenty feet deep, and piles could be driven only with great difficulty; no less than six were pulled out by the current. It was at first designed to scour out a deep pit by the use of wing dams, but before the plan could be carried into effect, the current, which is more variable at the site of this pier than at any other point on the line of the bridge, slackened to almost nothing, making wing dams wholly impracticable. A caisson of rectangular form was then built in position, 67 feet long, 30 feet wide, and 22 feet high. It was put together without spikes or pins, the planks being secured between cleets on the sides of the posts, and the whole caisson bound together by iron rods passing from the bottom of the sill to the top of the plate. It was proposed to sink this caisson about twenty feet, drive a pile foundation within it, cut off the piles at the level of the base of the caisson, build the pier on a suspended grillage, and lower it upon the piles. Then upon unscrewing the nuts and withdrawing the long rods, the caisson would fall in pieces and a riprap protection could be thrown close around the pier.

The slackening of the current was accompanied by a rapid deposit of sand, and before the caisson could be completed there remained but eighteen inches of water at the site of the pier; in eight days only, from the 18th to the 26th of September, a deposit twelve feet deep was formed. On the 26th of October the caisson was done, when it was tripped to the bottom by striking the braces which supported it. The change in the level of the river bed made a corresponding increase in the distance which the caisson must be sunk by excavation. This excavation was shortly begun by means of the steam siphon and hand dredge, and continued until the tools were transferred to Pier No. 3, the caisson having then been sunk about nine feet. During the winter an inner wall was completed within it, and the intermediate space, about five feet wide, was filled with stone and sand. An ice-breaker, formed of an inclined sycamore log and a fender of planked piles, was built above the pier site. In February a large dredge, with a steam-engine to drive it, was mounted upon the caisson; it was set in motion on the 13th of that month, and lowered the caisson a few feet. The rising water of the 8th of March produced a moderate scour, which aided the sinking; on the 17th the scour increased very rapidly on the south side, and the caisson began to tilt over; the next morning the water was found to be twenty-two feet deep there, while no corresponding wash had occurred on the north side. Under the combination of this undermining of the southern cutting edge, and the pressure of the sand against the north side, the caisson settled over till only the north-east corner remained above water. By hard work through the morning and dinner hour the machinery was removed and placed on boats. By 2 P. M. the whole caisson had disappeared; the weight of the sand and stone with which the walls were loaded, together with the external sand pressure, proved too great for so loose a structure; it broke in settling, and became a total wreck; a few of the timbers cleared themselves, and floated down stream, but the greater part of the wreck, being of green oak and covered with sand and stone, remained at the bottom of the river.

The loss of this caisson put an end to the work which had thus far been done on this foundation, making it necessary to start entirely anew. Moreover, the circumstances attending the wreck showed the exposure of this site to be so great that it was thought unwise to adhere to the plan of a pile foundation. The situation of this pier, between the edge of the sand bar and the low-water channel, exposes it to more frequent washes and deposits than have been observed elsewhere, while it is also liable to be subjected to the thrust of a heavy bank of sand on the north side, with no counterbalancing pressure on the south side, a danger from which the other piers are free. For these reasons it was determined to treat this as a channel foundation in preparing the new plan, and to extend the full-sized pier down to the rock. To avoid the difficulties of passing through the old wreck, which would have made it necessary to resort to the use of compressed air, and which it was feared would have delayed the completion of the bridge through another season, the location of the pier was shifted fifty feet to the south, reversing the distances between it and the two adjoining piers, and placing the long span, 250 feet, between Piers Nos. 4 and 5.

At the location now selected, the rock was assumed to be at an elevation of 55 feet and to be overlaid, during the best working season, with about 40 feet of sand, which would probably make it necessary to do some portions of the work in 50 feet of water. Borings taken indicated rock at 58 or 59, but were not wholly satisfactory. The methods by which the other deep sand foundations had been put in, though successful, had been very slow, and were likely to prove impracticable when the depth of sand became doubled, while, even if a bottomless caisson could be sunk to the depth now required, the season between two floods would be found too short to complete the work by putting in a subaqueous foundation of beton, 30 or 40 feet deep. For these reasons a plan was prepared resembling in many respects the process which was first introduced in founding the piers of the bridge over the Rhine at Kehl, and which has since been very generally employed by European engineers; in all previous works, however, the excavation has been made by laborers working in a pneumatic chamber—machinery, if used at all, serving only to remove the material which had first been handled by the men; but in these plans the machinery was so arranged as to be self-feeding, and the excavation was carried on without the use of compressed air. A pier of masonry was to be built in position above water, and sunk to the rock, by excavating the underlying sand with dredges working through wells left in the masonry, guiding the mass in its descent by suspension screws, and keeping the top of the masonry above the surface of the water by building on the successive courses as the sinking continued.

A caisson was designed which should serve as a support for the pier in its descent, and which, while of such form as should furnish the best facilities for excavation below, should bear, without yielding, the weight of 40 feet of masonry above, and the pressure of the sand and water against its sides. The construction of this caisson was begun on the 25th of June, 1868, on the north bank of the river, 400 yards below the bridge line. It measured 70 feet from nose to nose, 30 feet 6 inches in width, and 11 feet in height.* The sides were

---

* The plans of this caisson are given on Plate V.

built of square timber; the main sills were of oak, 15 inches square, of one piece from shoulder to shoulder; the seven succeeding courses were pine, 8 inches by 12, placed on edge, and the two upper timbers were oak, 12 inches square; a triangular piece of oak was placed below the main sill. The successive courses were pinned together with two-inch turned pins of oak, and bolted to uprights placed in the angles, and at intermediate distances along the sides; the outside was covered with two courses of three-inch oak plank, dressed in a planer to an even thickness, the planks of the inner course making an angle of 45° with the horizontal timbers, and those of the outer course being put on vertically, with the smooth side outwards. It was at first proposed to cover the whole with thin sheet iron to reduce the friction of the sand upon the sides; but experiments made to ascertain the coefficients of friction of sand against various substances, showed so slight a difference between iron and dressed oak, that the covering was not put on. Within this outer wall was placed a second wall inclined inwards; it was framed of oak timbers, 10 inches square, which rested upon the main sills and bore against a pair of 12 inch timbers, placed parallel with the upper timbers of the sides; this inclined wall was carried round the triangular ends, the framing being modified to accommodate the angles. Three braces, 15 inches square, were placed immediately above the main sills, extending across the caisson and bearing against the upright timbers; these served also as the bases of three ∨-shaped cross-walls, each formed of two equally inclined rows of oak sticks, 8 inches square, fitted into 15 inch timbers above; the lower angles of the cross-walls were formed by triangular pieces of oak, along the lower edges of which ran three iron rods, two inches in diameter, which passed through the main sills and tied the whole caisson together; each cross-wall was further strengthened by a truss built into the middle of it. The timbers of the inclined walls were thoroughly stayed by iron bolts binding them to the outer walls, and the cross-walls were strengthened by rods connecting their upper timbers; the interior framing of the starlings was secured by hanging it from a truss placed above, and the top of the caisson was tied across, by 2-inch rods placed at the shoulders, and by dovetailing the 15-inch cross-timbers into the sides. The whole interior frame was sheathed with 2-inch oak plank, but the spaces between the double walls were left entirely open above. The cutting edges of both main and cross-walls were protected by a covering of $\frac{3}{16}$ inch boiler plate, the plates being bent and cut to fit the angles and corners, riveted together and fastened on with wrought-iron spikes.

The combination of the $\bigvee$-shaped cross walls, with the inclined walls of the sides, divided the interior of the caisson into four bell-shaped chambers, the two central ones being nearly square, and those at the ends of pentagonal form, each having a rectangular opening above five feet and four inches by nine and a half feet. This form is one at once well suited to sustain the weight of super-posed masonry, and especially adapted to facilitate excavation. The caisson is thoroughly braced by the interior walls, and not encumbered with exposed brace timbers; the walls and edges are of such form as to act as wedges, which, under the weight of masonry, and by pressure above, feed the sand towards the centres of the chambers where the dredges work; while, as the cross-walls were placed thirty inches above the outer edge, a diver could have free access from chamber to chamber, should this be found necessary.

Twenty-four suspension rods, each twenty-four feet long and two inches and a half in diameter, with the upper end formed into an eye, were built into the walls. They were arranged in pairs, and passed through every square timber in the outer walls, taking hold with nut and washer on the under side of the main sills, the nut fitting into a square recess cut in the triangular stick below. Eighty $1\frac{1}{2}$-inch gas pipes were also placed in the caisson, arranged along the sides and cross-walls, and terminating in cast-iron nozzles immediately above the iron plating; they were intended for water-jet pipes, but the sand fed itself so well to the dredges that none of them except those in the angles were ever used. The whole planking was thoroughly caulked, and the interior coated with roofing pitch. A frame, provided with bolt holes, was carefully fitted into the rectangular opening above each chamber, and an accu-rate pattern taken, from which a cover could be made to fit this frame; so that in case extraordinary obstructions were encountered, the dredge could be with-drawn, the cover or trap placed in the frame, and bolted tight by a diver, converting the chamber into an air-tight caisson, when the obstructions could be removed by working in compressed air.

Under other circumstances it would have been preferred to build this caisson entirely of iron ; but the distance from adequate iron works, and the absence of boiler-makers and competent workmen, were unfavorable to doing so ; on the other hand, timber could be obtained without difficulty, and there was no scarcity of carpenters, so that it was thought best to build of wood, which involved much complicated detail and difficult framing.

The caisson was provided with a false bottom placed below the cutting edge, over which it fitted like the cover of a paper box, braced against the cross-walls, and secured by iron rods. Five launching ways were placed below, which were carried out into deep water on piles, and the completed caisson was lowered by jack-screws upon five flattened timbers, fitted with guides, and arranged to slide on the ways.

The first work done in the river was to drive a compact clump of anchor piles, one hundred feet above the proposed pier ; these were driven and protected by riprap before the June flood, but it was thought unwise to drive the false-work piles at that time, because, even if they should remain undisturbed by scour, they would inevitably collect a large amount of drift, which might form an obstacle in the way of sinking the pier scarcely less serious than the wreck of the old caisson. On the 9th of August this danger was past, and the driving of the false-work piles was begun. They were sixty in number, of which forty-eight, two to each end, were intended to carry the weight taken by the suspension screws, the other twelve serving only as supports for the false-works.* The two central piles on the lower end were not driven till after the caisson had been floated into place. The disturbances of the river made the driving of these piles less exact than it should have been, but the irregularities were not too great to be taken out in the platform above ; they were generally driven from twenty-five to thirty feet into the sand, some of them even reaching the rock.

The piles were cut off as soon as driven at an elevation of 106.5, a platform was built upon them, and the trusses were raised which were to carry the suspension screws. These trusses were seven in number, and proportioned to carry a safe load of 1,000 tons. Each end truss carried two suspension screws, and each of the intermediate trusses four, the screws being in pairs, and placed

---

* For arrangement of piles, see Plate V.

to correspond with the rods in the caisson. These screws were twenty-four feet long, ten of them three inches in diameter, and the other fourteen, which had been used in lowering the masonry of the Quincy Railroad bridge, two and a half inches. Ten additional screws of the same size as the latter were kept in reserve.

On the morning of the 21st of October the caisson was successfully launched and towed to the false-works. Two or three weeks previously a large flat-boat loaded with sand, in attempting to shoot the works, had struck against one of the upper piles and sunk; the wreck had caused a sand deposit at the pier site, so that, though there was plenty of water to float the caisson, which drew only three feet and a half, it could not be brought under the trusses without removing the suspension rods; they were, accordingly, unscrewed, taken out, and the caisson brought into position, when they were replaced and easily screwed into the nuts, which were held by the square recesses cut in the triangular timber below the sill. This was accomplished in a day, but the want of deep water proved a more serious obstacle in the way of removing the false bottom. It had first been proposed to sink the bottom by throwing in sand, water bein |already admitted above it, make fast to it with the steamboat, and pull it out be|ow; the depth of water proving insufficient for this, it had to be broken in pieces, and taken out in small parts, an operation which involved nearly two weeks' delay, and which, it was feared, would cause trouble by leaving unremoved fragments; an apprehension which fortunately proved groundless. A week later a sand bar, which had already been observed forming in front of the launching ways, had so much increased that it would have been impossible to launch the caisson, so that a tedious portage by land was narrowly escaped.

On the 11th of November, the work was begun of filling the spaces between the double walls of the caisson with beton, while the false-works were completed, and the machinery mounted as fast as could well be done.* The false-works were built with three floors; the lower one, intended for the use of carpenters and masons, was placed at an elevation of 108.7, and made a

---

* Full plans of these works and machinery are given on Plates IV. and V.

continuous platform extending on all sides of the caisson ; it was generally left open on all sides, but a small house was built at the south-east corner, in which a twenty-five horse-power engine and a donkey pump were placed ; a room was also enclosed in the middle of the south side for the use of the divers, where the air-pump and submarine apparatus was kept.   At the south-west corner a stair-case led to the second floor, which was placed on a level with the lower chords of the trusses.   This floor extended over the caisson, having four holes in it through which the dredges worked ; it was completely housed in, was provided with work benches, warmed by stoves, and contained the lamp-room and super-intendent's office.   At either end a staircase led to the third floor, a narrow platform, resting upon the upper chords of the trusses, where stood the four hand-crabs used in handling the dredges.

The excavating machinery consisted of four large dredges of the endless chain pattern.   They were mounted with vertical telescopic frames of wood, the lower tumbler being attached to a single frame, inclosed by a double frame which carried the upper tumbler ; the boxes of the upper tumbler were set on adjustable blocks.*   By this arrangement the dredges could be lengthened to suit the depth at which they were operating, the length being varied from 51 to 85 feet ; this was done by removing the bolts which united the two frames, putting in an additional length of dredge chain, with the proper number of buckets, and raising the outer frame till the length of the added chain was taken up ; the bolts were then replaced and such slack as might remain in the chain taken out with the adjusting screws.   The entire frames were raised and lowered, independently of this change in their length, by chains which passed through sheaves on the sides of the double frame, and were worked by the crabs on the upper floor.   Two of these dredges had originally been used on the Quincy bridge and were now rebuilt to adapt them to this work ; a third was similar, and had been made from the same patterns, though designed in the first instance for use at Kansas City, on the old No. 4 foundation.   These three dredges had square tumblers of cast-iron ; the links of the chains measured 22 inches between centres, and the buckets were bolted on every fourth link

---

* For plans and details of dredges, see Plates V. and VI. The plan on Plate V. shows the frame of shorter length than that actually used.

through holes drilled for the purpose. The other dredge was constructed especially for use on this pier; the tumblers were of hexagonal form, made of oak and bound with wrought-iron; the chain links were only 12 inches long between centres formed with upset ends ; the buckets, whose form was novel, were placed on every sixth link and held by the same pins by which the links were coupled. an arrangement relieving the links of any transverse strain, the merit of which was proved by the fact that the chain of this dredge never broke ; the hexagonal tumbler was also found to give a steadier motion than the square ones.*

A single line of shafting mounted on hangers attached to the trusses, and driven by the engine on the floor below, extended from the east end of the house till opposite the western dredge ; on this shaft were placed four pulleys, each arranged with clutch and lever, by which it could be thrown out of gear independently of the others, and the power was carried to the dredges by belts driven by these pulleys. The new dredge was mounted at the west end of the pier, and the dredge at ` e east end was also provided with buckets of the new pattern. These two dredges were worked through bevel gearing, the power being transmitted at any elevation by a pinion sliding on a vertical shaft ; the other two were driven more directly by the belts, which were kept tight under all elevations by a loaded tightener sliding in a vertical frame ; the latter arrangement proved the better one. Each dredge was completely boxed in between the second and third floors, to confine the splash, thus keeping the machinery and works upon the second floor dry and in good working order. The dredges discharged towards the north, the sand falling on inclined troughs which led to the lower platform, from which it was carried off in wheelbarrows, on runways built for the purpose, and deposited a hundred feet north of the works.

The machinery for handling stone was on the lower platform. It consisted of a railway and cars, the same which had been used at Pier No. 1, running along the west end of the works, and two travellers running lengthwise with the pier, between the wells and the sides of the caisson. A floating derrick was

---

* A patent for the improvements in these dredges was issued to the authors of this work, bearing date January 18, 1870.

moored on the south side of the works, by which the stones were lifted from the stone barges and placed on the car; the car was then pushed under one of the travellers, the stone raised by a hand-crab which was placed at the east end of the works till it cleared the car, and drawn forward till opposite the desired point by a steam crab, which was likewise at the east end of the works, and driven by the same engine which worked the dredges, both sets of machinery rarely being worked together. This apparatus was not mounted till the beton in the two lower sections of the caisson had all been put in.

Soon after the caisson was brought into position the rectangular openings into the lower chambers had been surmounted by timber boxes; this was continued from time to time as the sinking progressed, the successive sections of these well walls being made of such height as was found most convenient. When the hollow walls of the caisson had been filled with beton a second section was built above it; this section was an open frame structure, covered with three-inch oak plank dressed in a planer, and similar to the caissons used at Piers 1 and 3; the long sides were given a batter of one in sixteen, but the short sides of the starlings were built plumb; additional lengths were also put on the gas pipes. This section, like the lower one, was filled with beton, about one-half the full amount being put in before starting the machinery. The beton was mixed upon the platform, thrown at once into the caisson, and beaten down with a paving maul. It set rapidly, forming a satisfactory compound; the caisson thus became merely the wooden covering of a single artificial stone or monolith, of the form most convenient for the work, and which carried the masonry of the pier above.

On the 11th of December the ice closed at the bridge line, and the river froze across. A week later the ice, which was still thin, began to rot rapidly under a strong sun, and on the 19th it broke up and went out. No serious damage was done, but a large sheet of ice, jamming above the draw rest, forced inwards the ice along the north shore, which, swinging on a pivot, about the anchor piles above the works, tore out two piles on the north-west corner of the false-works of this pier, the injury being done at one of those points where the exposure was supposed to be least. The damage was soon repaired; one of the piles had only been bent over, and was drawn back into place; the other, the corner pile, was destroyed, but the platform was made secure by bracing below.

On the 28th of December the machinery was started; a few unimportant changes were found desirable, but its performance was, on the whole, very satisfactory. For the first week it was driven only by day, while the forces were being organized and drilled to their work. On Monday, the 4th of January, two gangs were put on, and the work proceeded both night and day. Each gang had a superintendent at its head, Mr. Tomlinson taking the day, and Mr. Bostwick the night shift; a master of machinery had general charge of the four dredges, while two mechanics were assigned to the care of each dredge; an engineman and fireman tended the engine on the lower floor, another man was given special charge of the donkey pump, and a spare machinist was employed upon odd jobs; a large gang of laborers completed this force; all the laborers worked under one foreman, and the majority of them were employed in wheeling off the sand, but twelve men were detailed to work the crabs on the top floor, and a few more to tend the suspension screws, while it occasionally became necessary to call in the entire force for the latter work. The same force was, of course, duplicated for the second shift; each gang worked from seven to seven o'clock, the day gang being allowed an hour at noon for dinner, and the night gang being furnished with hot coffee at midnight.

Eight vertical rods, graduated into feet and tenths, were fastened on the sides of the caisson, one at each end and shoulder, and one in the middle of each long side; they served as gauges to measure the descent, eight blocks placed on the platform opposite them, at an elevation of 109, answering as reading fingers; the gauge at the west nose was numbered one, that at the south-west shoulder, two, and so on continuously around the pier. The dredges were also numbered from one to four—the new dredge at the west end being number one. A full journal of the progress of the sinking was kept by the superintendent, from which a set of tables, illustrating the behavior of the pier and conditions of the sinking, were prepared. These tables, which give the best illustration of the actual working of the plan, are printed in Appendix E.; they contain a statement of: 1st. The number of hours' work performed by each dredge, with the estimated daily excavation. 2d. The readings of the gauges, daily progress, and average elevation of the cutting edge. 3d. The soundings opposite each gauge, and average elevation of the sand surrounding the pier. 4th. The displacement and the actual and effective weights of the pier. 5th. The area of the surface in contact with the sand, and the effective weight for each square foot of such surface in contact, with estimated friction.

The material dredged was at first a soft sticky silt, which could be handled only in connection with a large amount of water in the form of a thin, flowing mud. The work was conducted very carefully, the gauges were constantly watched, and the screws were tended continually; with these precautions little difficulty was experienced in keeping the pier true; after it had been sunk ten or twelve feet the surrounding sand answered as a guide, and less care was required to regulate the descent. Owing to the weight of the pier and the care with which the machinery had been arranged, the sinking proceeded at a very much more rapid rate than had yet been accomplished with the bottomless caissons, and exceeded the expectations of the engineers. On the 6th of January, only two days after both shifts of men had been put on, the work had to be suspended, because the beton could not be put in fast enough to keep pace with the descent, and from this time forward the chief difficulty lay in building up the pier rather than in sinking it. The water jets were found to be of less service than had been anticipated, the wedge-shaped edges feeding the sand to the dredges without their assistance; streams of water were occasionally passed through the pipes at the nose and shoulders, and all the outside pipes were lengthened as the height of the caisson was increased, but those in the cross-walls were allowed to be buried up in the beton.

On the 7th the machinery had to be stopped again, and it remained idle nearly a week; on the 8th the beton was nearly all in, reaching to the top of the second section. A third section had meanwhile been added, twelve feet high, the end walls of which were at first made only one-half this height, to facilitate handling the stone. On the 9th the river rose about a foot, causing a strong current on the south side of the works, which was found to have increased the depth of water from nine to seventeen feet, so that the pier began to settle over slightly, till held by the suspension screws; one hundred and fifty gunny bags were filled with sand and thrown overboard among the piles and along the side of the caisson, which suspended the scour.

On the 13th the masons began work, laying the first course of stones on the hardened surface of the beton ; in the evening of the same day the dredges were again set in motion, and the work of sinking resumed. The following day the river began to rise again, repeating the scour of the preceding week ; the wash was again restrained by the use of sand bags, over five hundred of which were thrown around the works on this and the two succeeding days. This method of protection was found effective, while it was free from the objections which prevented the use of rip rap ; if stones had been thrown around the pier it was feared that they might work under the edge of the caisson and obstruct the descent ; the sand bags might also work under the edge, but their soft and yielding nature would prevent their doing harm ; some of them did actually find their way inside of the caisson, and one was brought up in a dredge bucket uninjured.

It being found impossible to lay masonry as fast as the dredges could sink the caisson, the plan was adopted of running the machinery only by night, and giving the masons every convenience for work by day. Mr. Tomlinson then took charge of the night shift, and the pier was sunk for the remaining distance under his directions. The masons were often unable to do more than set the face stones of a course, together with a few of the heavier pieces of backing, in a day, in which case the night force would be employed during the first hours of their shift in backing up with beton. The material excavated had changed to a coarse sand which was easily handled, each dredge throwing six full buckets in a minute ; the pier also settled more rapidly than hitherto, sinking five inches in an hour when everything was working well. The lower platform and the second floor were lighted by locomotive head-lights, which threw a strong glare over the works and men, and a visit to the pier late in the evening, when the machinery was all working to its best advantage, and half an hour showed a decided settlement, became a very interesting thing.

On the 19th, a pile top was found buried in the sand below dredge No. 2, which was secured by a diver and drawn out with little trouble. The next day some timber, supposed at first to be the branches of a large snag or tree, was discovered under the lower end of the caisson. An additional diver was sent for, and after a few days' delay the log was cut through and drawn out, when it was found to be a broken pile, probably belonging to the works of the wrecked foundation. Another old pile was found near it, which extended from outside of the caisson nearly to the centre of the eastern chamber, passing under the cutting edge ; a line was made fast to it and attached through a set of falls to the upper false-works, and held in this manner while the sinking proceeded ; on the 1st of February, this pile broke off under the cutting edge and was drawn up through the well hole ; it proved to be a stout hickory stick, nearly a foot in diameter, and showed a rough broom-like fracture ; it had been carried down with the caisson several feet before breaking, and the outside portion still remained under the edge, where it was found by a diver when the caisson had nearly reached the rock. While the divers were at work upon these sticks, it became necessary to jet away the sand around them, thus forming a cavity close to the edge of the caisson ; in two or three instances this caused sand slides, the sand suddenly caving in, filling up the cavity and raising the water in the wells ; at one time the water in the wells was raised three feet above the level of the river, when the soundings showed a hole ten feet deep outside the caisson, over the point where the slide occurred ; this, however, was soon filled up by caving in and by fresh deposits.

On the 3d of February, the masonry was finished to the top of the sill of the fourth section, which had now been added, or thirty-nine feet above the cutting edge ; as this was less than two feet below the point at which the ice-breaker courses were to be started, it was thought best to lay no more masonry till a permanent bearing had been reached upon the rock. Borings recently taken had found the rock at an elevation of 56.6, though the auger had apparently been disturbed by loose stones three or four feet before it reached that depth. In the evening of the 4th the pier settled rapidly ; the machinery had never worked better, and six inches descent was noticed in an hour ; but at mid-night it came against some hard substance and almost stopped. The diver at first reported rock, but the pier went down nine inches during the next three days, and though the dredges threw out a large number of loose stones, the obstruction was found to be a mass of clay under the south edge ; the upper section was filled with sand, and under the pressure of this additional weight, seventeen inches more descent was obtained. It was evident, however, that the bed rock was covered with about three feet of loose stones mixed with a moderate quantity of stiff blue clay ; the foundation would probably have been perfectly safe if kept where it was, but it was still thought best to place it directly upon the rock.

An additional number of divers were engaged, and on the 16th of February, a force of eight divers with four air-pumps and the proper complement of tenders, was ready for the work ; they were divided into two gangs, and the work was prosecuted both nigh and day, one man working in each chamber. The depth of water in the wells was about fifty feet, and to render the work less burdensome, the water was warmed by sending steam down the water jet pipes. The stones were removed singly from under the edge, piled up in the centre of the wells, and placed in the dredge buckets ; the dredges were worked for a short time after the divers had come up, bringing up the smaller stones ; the largest rocks were left below. The stones were of all sizes, from small pebbles to boulders containing two or three cubic feet ; the larger ones were mostly of limestone, and showed few or no signs of wear ; the smaller pebbles were well rounded, and of diverse geological character, presenting a strange collection of the different formations found on the eastern slope of the Rocky Mountains ; sandstone, granite, moss agates, and many other minerals were mixed in wild confusion, while bits of water-charred wood, reduced almost to pure coal, and several varieties of teeth, were found among them ; an Indian arrow head was also picked out of the lot.

On the 10th of March the rock was reached, at the elevation of 56.6. A hole was drilled into it five feet, as had been done at the three other channel foundations, and no sign of any flaw or weakness discovered. A row of bags, filled with freshly mixed beton, was placed around the edge, as had been already done at two of the other foundations, and the dredges were removed and the wells filled up with beton, laid under water, with the same boxes that had previously been used at Pier No. 2. Divers were still employed, to make sure that the beton filled up the whole space of the lower chambers, packing well in towards the edges, and covering the boulders which had been left piled in the centre ; the sand was thrown out from above the masonry, and the upper sections of the well walls were torn away, to secure a good bond between the masonry and the filling of beton.

The layer of boulders had been the cause of considerable delay, while it had also been productive of some additional expense ; but the character of the larger stones, which, by their roughness, showed that they were seldom, if ever, disturbed by the water, indicated the perfect security of a foundation put in at this depth ; and the mere presence of such material was equivalent to three feet of riprap protection around the base of the pier.

The false-works were stripped, the trusses taken down, and on the 26th of March nothing remained above the lower platform. A fifth section was added to the caisson as a security against any rise in the river ; a derrick was mounted on the platform on the north side of the pier ; on the 2d day of April the laying of masonry was resumed and the pier was built up at once.

This foundation, which from its situation might fairly be regarded as much the most difficult on the work, became, in its final execution, the most successful of all, and was put down in a less time than was consumed on any other deep foundation. The plan here adopted is believed to admit of wide application ; and, while it is more expensive than the simple foundations which are used in ordinary streams, it becomes a cheap method of founding in deep and unstable bottoms. By slight modifications it can be combined with the pneumatic process, in such a way as to allow extraordinary obstacles to be removed by men, while the entire sand excavation is made by machinery. It is also applicable to foundations of extraordinary depth, where the pneumatic process must fail from the inability of the men to stand the air pressure ; it could be carried to a depth double that to which pneumatic tubes or caissons have been sunk, with the occasional use of the air chamber for a very short time ; and if this be entirely dispensed with, it may even be extended to a depth of several hundred feet in clean sand, or with machinery sufficiently heavy to remove obstacles.*

---

* A patent for this method of founding has been applied for by the authors of this volume.

PIER No. 5.

Although the loss of the first caisson made Pier No. 4 practically the last foundation begun, No. 5 was the pier on which the original work was latest taken in hand. The erection of the caisson was begun on the 20th of January, 1868 ; it was built in position on a ·y sand bar, and in general form and construction it was similar to the caisson used for Piers 1 and 3 ; the upper end was built entirely of square timber, the cutting edge was made by cutting off the vertical planking with a bevel, and the spaces between the timbers were filled with rubble masonry. A cluster of anchor piles was driven fifty feet from the upper nose, and the false-works, with their supporting piles, were placed inside of the caisson. The object of this device was to arrange the works in that form best adapted to withstand the washing of the spring floods, but the site of the pier proved to be beyond the range of the scour and the precaution a needless one. The lower section of the caisson was finished on the 13th of February ; it was attached to 12 long screws, eight of which had been previously used at Piers 1 and 3, raised from the blocking it was built upon, and lowered to the ground.*

After the wreck of the old caisson at Pier No. 4, the machinery which had been in use there was set up at No. 5, and dredging was begun on the 26th day of March. The small dredge used at Pier No. 3, and a second large dredge which had been used on the Quincy bridge, were shortly added to this equipment ; the three dredges were mounted above the caisson, the small dredge being placed in the middle, and one of the large ones at each end, all discharging towards the south ; they were driven by a single 25-horse power engine, the one subsequently used to drive the whole machinery at Pier No. 4, which stood on a platform built on piles directly north of the caisson. The sinking went on very slowly ; the sand did not flow readily to the dredges, and the amount of excavation was greatly in excess of the displacement of the caisson. The dredges often dug out holes, working down 8 or 10 feet below the base of the

---

* For plan of this caisson see Plate VI.

caisson, while the material under the edge was but slightly disturbed ; the sand round these holes would then fall in, bringing with it considerable quantities of sand from the outside ; but the slides seldom extended far enough along the edge to cause any material settlement in the caisson. A water jet, attached to a long piece of gas pipe, and handled with lines from above, was passed around the edge of the caisson on the inside, which helped greatly to clear away any interior sand bearing, to level off the material, feed the dredges, and let the caisson down ; the dredges were usually run during the greater part of the day, excavating a moderate quantity of silt and sand, but causing no perceptible descent ; they were then stopped, and the pumps started, when an hour or two of jetting would be accompanied by a few inches settlement. The material excavated was at first a fine silt; but as the depth increased it changed to a coarse sand, in which were found occasional masses of clay, and a few stones. In one instance the caisson reached what appeared to be a thin continuous layer of clay, which the dredges worked through without producing any general effect, and which had to be cut in pieces with chisels from above.

On the 27th of April a second section was added to the caisson, which was loaded with a wall of rubble masonry, between the timbers, like the first section. On the 5th of May the machinery was stopped, the dredges rearranged by transferring their support to the top of the second section, and started again on the 14th. To secure additional weight, a box five feet by four was built around the caisson, supported by brackets on the outside and filled with sand. On the 5th of July a third section was added, and the sand boxes were raised to prevent their taking a bearing on the outside sand. About this time the summer flood covered the sand bar, so that these works could only be reached in boats. The level of the bar, between this pier and No. 7, was raised by a fresh deposit of two or three feet of silt, and its general line was extended about twenty yards towards the south, carrying it beyond the site of No. 5, but no material changes occurred immediately around the caisson; the current caused by the obstructions of the work, and the frequent sand-slides about it, prevented the accumulation of any deposit, leaving the works in a little bay by themselves, and the operations were not in the least disturbed by the high water.

On the 15th of August the base of the caisson reached an elevation of 80,

30 feet below the level of the adjacent sand bar ; the character of the material in which the dredges were working, it being more of the nature of a gravel than of a fine silt or sand, showed that scour rarely reached below this depth, and that it could be confined to a higher level by the moderate use of riprap. For these reasons it was determined to stop the excavation here ; the machinery was removed, the upper section taken off, and two pile-drivers were mounted on a timber scaffolding above the caisson ; a steam-engine on the shore supplied the driving power, and on the 10th of September pile-driving was begun.

The piles were provided with cast-iron shoes ; each shoe was cast hollow, with a round hole at the point, and fastened to the pile by four wrought-iron straps moulded into the casting ; a groove was cut with a broad axe in the side of the pile, and in it was placed a gas-pipe, the lower end of which terminated in a hole in the head of the shoe ; the gas-pipe was connected with a donkey pump by a flexible hose.* The pile, with its attachments, was placed in the leaders of a common pile-driver, and the pump started, which forced a stream of water directly out of the point of the pile ; the hammer, weighing 2,200 pounds, was lowered gently upon the head of the pile, which would settle from 12 to 15 feet as the sand around it was loosened by the jet ; the hammer was then raised and a few gentle blows struck, after which the pile was driven by repeated hammering as far as length would permit ; an iron-bound follower of oak was then placed upon it, and the driving continued till no settlement whatever could be observed. One driver was placed at each end of the caisson, and the two worked forward till they met at the centre ; the driving was very slow, it usually requiring 24 hours to set a pile and drive it home ; the piles were placed and driven as far as possible by day, and then followed down by night. Borings had found rock at an elevation of 48.5, 31.5 feet below the base of the caisson, and a record was kept by the foreman of the depth attained by each pile ; nearly all of those first driven reached the rock, but as the work proceeded the sand became more compact, and it was found impossible to force down the last 40 piles to that depth. One hundred and forty-four bearing piles were driven, some of them being struck over 1,000 blows, besides about a dozen piles along the sides of the caisson to serve as stiffeners.

---

* This shoe is shown on Plate VI.

The pile-driving was completed on the 1st of December. The piles were then cut off under water at an elevation of 89.9 with a circular saw suspended from a movable frame and worked by six men. The space amongst the piles, above the sand, was filled with beton, which served to tie the piles together, and, being enclosed by the caisson, would of itself form a sufficient foundation, if not undermined by the scour. The pit was then pumped dry and the piles were capped with flattened sycamore sticks, on which was laid a second course of timber of the same kind, the hollow spaces being all filled with beton; on this the masonry was started on the 14th of January, at an elevation of 91.1.

During the low-water season, after the completion of the bridge, a large quantity of heavy stones which had been left over from the masonry, were collected and placed around the base of this pier, forming the foundation of a substantial riprap protection.

PIER No. 6.

A few piles were driven on the proposed site of this pier in the spring of 1867, the locations of the piers being then so arranged as to divide the distance of 577 feet between Piers 4 and 7 into three equal parts. After the summer flood these few piles were found to have collected a quantity of drift and roots, which was partly buried in the sand bar, and could be removed only with difficulty; the location of the pier was accordingly moved 15 feet to the north, or to a point 400 feet from the old site of Pier No. 4, and these obstructions were avoided.

On the 3d of September, 1867, work was begun by driving an enclosure of sheet piling. The sheet piles consisted of pieces of four inch oak plank, the edges grooved and pointed to fit one another, and the lower ends sharpened to a feather edge. They were driven between wales, which were bolted on square timbers previously driven, a wooden follower being interposed between the iron hammer of the driving engine and the planks.* On the 28th of September the enclosure was completed, when a pit was dug with shovels within the piling; the excavation was carried to a depth of 15 feet below the sand bar, reaching

---

* The plan of this enclosure and a detail plan of the form of sheet piles are given on Plate VI.

an elevation of 94 ; the sheet piles were driven down as the work proceeded, and the water kept out by a steam-pump.

Three pile-drivers were then mounted over the pit, and the bearing piles of the foundation driven ; these were 90 in number, and their average penetration was a little more than 30 feet. The drivers were at first worked by horse-power, but this was found unprofitably slow, and two of them were afterwards worked by steam. This done, the pit was again pumped out on the 16th of December, and the excavation carried a foot and a-half lower. The sand was dug away from the outside of the sheet piling for eight or ten feet, to relieve the pressure, but it was found difficult to excavate much faster than it flowed in on the inside ; the water also came in in such quantities that the two centrifugal pumps were required to keep it down. The piles were then cut off with axes and the heads worked smooth ; they were capped with flattened sycamore sticks, on which a second course of timber was laid, which was planked with four inch oak plank, finishing at an elevation of 94.7. On this the masonry of the pier was begun on the 3d of January, 1868. The pier was built up at once, the pit around it being filled with riprap.

PIER No. 7.

As early as February, 1867, an excavation was made in the side of the bank, at the site of this pier, and in this the foundation piles were driven at once. They were 73 in number, and their average penetration 27 feet ; their driving occupied about a month, from the 27th of February to the 27th of March. The excavation was then resumed around the piles, but after a week's work the men were driven out of the pit by the rising water of the April flood. The water continued too high for the work to be resumed till after the summer floods ; when it fell in August no trace of the foundation could be seen, the piles having been completely covered by the deposit of sand. The pier was therefore located anew, an excavation made at the site, and the buried piles dug out, in nowise injured by their premature inhumation. A rough enclosure of sheet piling was driven, and the excavation continued below ; the water came in rapidly, working its way through the porous soil, and making the excavation exceedingly troublesome.   The piles were cut off and capped, and the masonry started on the 1st of October, at an elevation of 101.1.   The earth was afterwards replaced around the pier and carefully paved with rubble, effectually excluding the timber-work from the air, while the moist nature of the soil renders it as imperishable as if perpetually below the level of the water in the river.

CHAPTER IV.

MASONRY.

THE contract for the masonry of the bridge was originally let to Messrs. Vipond and Walker, of Kansas City. On the 1st of July, 1867, Mr. Walker retired from the firm, and his place was supplied by Mr. J. H. Burns; on the 25th of November, in the same year, Mr. Vipond died, and the work under the contract was completed by Mr. Burns.

By the terms of the contract the Company agreed to furnish all derricks required for handling stone, both at the quarries and on the river, and also stone-boats for transporting stone to the pier sites, and the use of their steamboat to tow these boats to the points desired. The contractors were to furnish all smaller tools, to provide power for the derricks, and to keep derricks and boats in repair while in their use. The cement was to be purchased by the Company and charged to the contractor at cost.

The stone used was limestone, the greater part of which was quarried in the bluffs south-west of the city, and within three miles of the bridge site; a quarry was also opened on the north side of the river, from which a portion of the stone used in Piers 5 and 6 was obtained. Several varieties of stone were worked, the best of which was a compact blue limestone, of nearly uniform color, found in continuous layers varying from 15 inches to two feet in thickness. As this stone could not be obtained in large quantities without very expensive stripping, its use was confined to the ashlar work of the upper parts of the piers; the whole of the piers, below the top of the ice-breakers, was built of a more coarsely grained stone, of a white or gray color, which worked into thicker courses than the blue stone, and which was used for backing throughout. The stone was quarried in the summer and early autumn of 1867, so as to allow a sufficient time for seasoning; it was found in general to stand the frost well,

10

with the exception of one lot of very heavy stones from a single quarry; these were badly broken by the first heavy frost of November in that year, and the products of that quarry were condemned for dimension work above low water.

The specifications required the work to consist of the best description of rock-range work, the face stones to be cut, squared, and bedded with one-quarter inch joints, and with the vertical joints cut back at least nine inches from the face; the ice-breaker faces were to be cut smooth, and drafts cut on all angles; the shoulders and corners were to be trimmed so as to have no projection exceeding one inch and a quarter, while no projection exceeding four inches was to be allowed on any part of the pier. The whole size of the top of each pier was finished smooth, and the stone bush-hammered, the face of the coping being also trimmed almost smooth. The face stones were fastened together by iron cramps of inch round iron, as high as the top of the ice-breakers, and this system of dowelling was continued at the shoulders up to the overhanging courses, where it was again extended to the whole face. The backing was formed of heavy uncut stone, laid in full mortar beds, the crevices being filled with smaller stones laid also in mortar. The whole amount of masonry was laid in hydraulic mortar, the usual proportions of the mixture being two parts of sand to one of cement; in the upper courses, which are rarely or never exposed to the water, this mortar was mixed with a paste of fat lime. The hydraulic cement was of the well-known Louisville manufacture, the greater part being purchased from the Falls City Cement Company.

The masonry contract included the beton used and the riprap thrown around the piers, though not the river protection above the bridge. The beton was formed of broken limestone, sand, and cement, the proportions varying with the purpose for which it was used. The stone was broken by hand into pieces that would pass through a three inch ring. The method of preparing the beton for use, was to mix the mortar separately in a grout box, and then pour it with pails over the stone, which had previously been spread evenly over the floor and moistened with water to remove all dirt; the mixture was then rapidly turned over with shovels and deposited at once in its place. If placed at once under water, it was lowered carefully in boxes of the patterns already described. The beton used at Pier No. 3, was formed of six parts of stone to three of sand and two of cement, a slightly larger proportion of cement being used in the first few feet. It was found, however, that so large a measure of sand was not favorable to the rapid setting which is important when the beton is exposed to the water from the very first, and in the subsequent foundations this proportion was changed ; the beton used at Pier No. 4, consisted of eight parts of stone to two of sand and three of cement ; that used at Pier No. 2 had nearly the same constitution. In laying beton under water considerable inconvenience was found from the *laitance* which formed in large quantities, especially if the mortar had been mixed too thin, or if the water was very cold ; it was sometimes necessary to suspend the work for a day or two, and pump out the *laitance*, but it was generally found sufficient to pump for a few hours every night, though when working in this manner care had to be taken to avoid washing the beton before it had set.

Both masonry and beton were laid in extremely cold weather, the use of hot sand and water being found to make this perfectly practicable. The sand was heated in large sheet-iron braziers, and the water warmed in cast-iron kettles, one of each being found sufficient to supply the force working on a pier. The heat, which was thus artificially given to the mortar, hastened its setting, causing this to take place before the mass had cooled enough to make freezing possible.

The form of pier adopted is somewhat unique, and was selected from the advantages it was thought to offer in combining a roomy bridge-seat with a slender and graceful pier. The accompanying lithograph, representing Pier No. 1, and taken soon after its completion, shows the general form of the oblong piers.* These piers are built with a side batter of three-quarters of an inch to a foot, or 1 in 16, and the same on the starling ; the ice-breakers have a batter of six inches in a foot, or one in two, giving to the cutting edge of the nose a retreat of eight inches and a-half in each vertical foot. The angle made by the two faces of the starling, measured on a horizontal plane, is a right angle. The ice-breakers finish at an elevation of 116, this being considered the greatest height at which the ice will ever move in the river ; the height at

---

* The full plans of the several pieces of masonry are given on Plate VII.

which they begin varies from 97.5 to 100, the latter being the height fixed by the original plans, and adopted on Piers 1 and 6, but subsequently changed on observing the exceedingly low stage at which the ice went out in 1868. The ice-breaker nose is protected by a heavy plate of cast-iron, and the shoulders are carefully dressed to a curve; this cutting being done after the stones were laid in the pier. The overhang of the cornice is one foot on each side of the pier, and two feet on the starlings, making two feet and ten inches on the angle of the nose. All the oblong piers have the same total length, the difference being in their thickness. Piers 1 and 3 finish eight feet thick at the neck, and ten feet broad on top; Piers 4, 5, and 6 finish seven feet thick at the neck, and nine on top; and Pier No. 7, six feet at the neck and eight on top. As Pier No. 7 is situated within the line of the shore it was built without an ice-breaker. All of those piers finish at an elevation of 145.6.

The pivot-pier is of circular form, 29 feet in diameter, and built plumb without a batter. The cornice has an overhang of one foot and a-half, making the diameter on top 32 feet. This pier finishes four inches higher in the centre than on the circumference, this difference being made to accommodate the dimensions of the turn-table; its elevation on the outside is 142.24. The two pillars on the bank are of square section, measuring seven feet and eight inches on top and finishing at the same height as the oblong piers. The south abutment is built with its ends parallel to the trusses of the 68 foot span, and finishes at an elevation of 147.10.

These sizes make the actual clear openings of the draw 160.38 feet, at an elevation of 100, the lowest navigable stage of water; 162.8 feet at the neck of the piers, where the piers are narrowest, and 160.25 feet between the copings.

The first stone laid was in the south abutment, on the 21st of August, 1867; work was suspended here during the building of Pier No. 1, and the abutment was not completed till the latter part of December. The pillars were begun in December immediately after the completion of the abutment, and finished in the following month. The time occupied in building the several piers is shown by the following table:

Pier No. 1, First stone set October 16, 1867. Pier completed November 30, 1867.

"	2,	"	February 20, 1869.	"	April 21, 1869.
"	3,	"	May 28, 1868.	"	September 14, 1868.
"	4,	"	April 2, 1869.	"	May 5, 1869.
"	5,	"	January 14, 1869.	"	March 9, 1869.
"	6,	"	January 3, 1868.	"	February 14, 1868.
"	7,	"	October 1, 1867.	"	November 15, 1867.

This does not include the masonry laid in Pier No. 4 during the sinking of the foundation.

The masonry of Piers 1, 2, and 3 was laid entirely with floating derricks, which were also used to some extent at Pier No. 4.* These tools were found among the most serviceable parts of the bridge outfit, and admirably adapted for use on the Missouri river; by chaining them to the sides of the caissons, or lashing to them spars which rested on the bed of the river, all difficulty from lurching under heavy loads was obviated. The three northern piers, as well as the south abutment and pillars, were built with land derricks of the ordinary pattern.

The amount of masonry and beton in the several piers is given below. These quantities are the actual amounts returned in the contractor's final estimates, but include only what forms a part of the permanent work, taking no account of the rubble masonry used in weighting the caissons for Piers 2, 3, and 5, nor of the beton placed in the draw rests :

South Abutment,...... 197 cubic yards Masonry.

Pillars,.............. 90 " "

Pier No. 1,............1,234 " "

"	2,............1,199	"	"	767 cubic yards Beton.
"	3,............ 873	"	"	833 " "
"	4,............1,109	"	"	1,169 " "
"	5,............ 820	"	"	333 " "
"	6,............ 656	"	"	
"	7,............ 434	"	"	

---

* Plate VIII.

CHAPTER V.

SUPERSTRUCTURE.

In the early part of August, 1867, letters were sent to a number of prominent American bridge-builders, inviting proposals for the superstructure of the Kansas City Bridge. These letters were accompanied by sets of specifications of general character, which were intended to serve rather as an indication of the quality of bridge wanted, than to contain the precise requirements of a contract. The lengths of the several spans, and the uses for which the bridge was building, were given in these specifications; they also stated that it was designed to build the draw entirely of iron, and the fixed spans of a combination of iron and wood, the latter material being used used only to resist compressive strains; the moving loads to be assumed in the calculations were specified, as well as the strains to which the iron might be subjected, and the factor of safety to be used in the wooden parts. The builders, however, were invited to propose any form of truss which they might select, submitting plans of the same if novel, and to suggest such departure from these specifications as might in their judgment seem wise, with the reasons for the change, and a statement of the benefit resulting therefrom. At the same time a set of plans for the fixed spans was prepared by Mr. Tomlinson, under the direction of the chief engineer, which were to be adopted only if, on a fair examination, they were found to be preferable to those submitted by outside parties. It had been intended to prepare plans for the draw as well, but in consequence of the mass of detail which this would involve, and the shortness of the time, it was found impossible to do so.

Nine sets of proposals were received from five different parties, two being on the common Howe truss plan, with both chords of wood; of the other plans, three were adaptations of the Pratt truss, one being entirely of iron, and the remaining four were respectively examples of the Post, the double and the single triangular, and the Fisk suspension trusses. On the 30th of October the contract was let to the Keystone Bridge Company, of Pittsburg, Pa.; the fixed spans were to be built according to the plans supplied by the chief engineer, the iron in them being paid for by the pound, and the timber by the foot; the draw was to be built, according to the contractors' design, for a fixed sum; subject, however, to such alterations as might be suggested by the chief engineer, the Company to have the benefit of any saving which might result from such changes, and to pay any extra cost which they might involve. Under this provision certain changes were suggested in the depth of the truss and arrangement of panels, which resulted in a material reduction of the cost. By a subsequent arrangement a pony truss of wrought-iron, made by the contractors from their own designs, was built, in place of the composite structure proposed by the engineer for the shore span of 66 feet.

The general design of the fixed spans is that of a double triangular truss or trellis girder, in which the top chord, posts, and braces are of wood, and the other members of wrought-iron, cast-iron being used in the details and connections. This combination, which has been used as yet only to a limited extent, is believed to overcome the most objectionable features of a wooden bridge, avoiding the wasteful connections which accompany the use of wood in tension, and disposing of the bulk of the perishable material in places where it can easily be protected; besides this, the character of the butt-joint connections, used to take compression, is such, that worn out parts can be removed and replaced by others without disturbing the remaining parts of the structure; it is also possible to replace the wooden parts by iron, and thus gradually convert the bridge into an iron structure without the expense of false-works or the intermission of traffic. The braces, which are always open to the air on all sides, are exposed to moisture only during the actual prevalence of a storm, and would therefore be well protected by a thorough coating of paint. The top chord can be covered in, and thereby thoroughly protected from the weather, without perceptibly increasing the wind surface of the bridge. The only danger to which such a bridge can be exposed is that of fire, and if the wood-work be painted throughout with mineral paint, and a watch kept, which is always necessary at Kansas City, men being constantly needed to tend the draw, and collect tolls, this danger is reduced to almost nothing.

The trusses of the five fixed spans measure respectively 130, 198, 248, 198, and 176 feet, the difference between these distances and the lengths of spans, given in the preceding chapters, being the allowance made for pedestals, wall-plates, and clearance room. The two shortest of these have straight parallel chords, the depth of truss being 22 feet; the same depth is retained at the ends of the larger spans, but in them the upper chord is arched so as to increase the central depth to one-eighth of the length, the inclination of the braces being kept nearly constant by varying the lengths of the panels. The upper chord of the 130 foot span is formed of three pieces, packed in the usual manner; in the other spans the chord is of five parts, and supplemented at the centre by a sub-chord of two parts. The lower chords are of wrought-iron upset links with pin connections, made under the Linville and Piper patent. The end posts and braces bear upon a cast-iron pedestal, which rests on a wall-plate likewise of cast-iron, carefully fitted to the masonry, and well bedded with mortar; at one end of each span a set of rollers is placed between the pedestal and the wall-plate. In place of the ordinary square ends the braces are cut with two end faces, which make an obtuse angle with one another, and the angle blocks are cast to correspond; this device makes it impossible for a brace to slip upon its bearing. The ties are of square iron, with a welded loop at the lower end, passing around the chord pin, and a screw cut on the upper end, which is previously upset, so as to leave an equivalent area after the cutting of the screw. In the 130 foot span both the main and counter braces are single, the counters bearing upon cast-iron brackets placed on the sides of the main braces; the main and counter ties are in pairs running along the sides of the braces. In the other spans the main braces are in pairs, and the counters, which are single, pass between them. In the 176 foot span both sets of ties are in pairs, the main ties passing outside of the main braces, and the counter ties between the main and counter braces. The arrangment of ties is the same in the central panels of the 198 and 248 foot spans, but in the panels near the ends there are four main ties, two passing outside the main braces, and two between them and the counter. In all the spans the counter ties are carried only so far as the stiffness requires, but a counter brace is placed in every panel to take a bearing in screwing up the main ties.

The most novel detail of this truss is the top angle block; this is of cast-iron and formed of three forms of castings.* The respective parts are: *First*, the angle-block proper or brace-bearing, which is placed below the chord, and receives the ends of the braces in the central panels of the four largest spans; this is cast with extended ends, to form a connection with the sub-chord. *Second*, the keys, which pass through the chord in much the same manner as ordinary packing blocks; they are cast hollow, in as many parts as there are spaces between the chord timbers, and with side plates to receive the ends of the timbers whenever a joint is broken. *Third*, the washer plates, which rest on the top of the chord and carry the nuts of the ties; the plates for the main and counter ties are cast separate. The brace-bearings are cast with flanges extending their whole length, which fit into grooves cut in the chords, and bear against the cast-iron keys; the ties pass through the hollow keys, nowhere coming in contact with the wood of the chords. As the ties take hold of the washer-plates above, and the braces rest against the bearing below, both of which bear upon the keys, the strain is distributed, from the first, through the whole section of chord, instead of being thrown entirely upon one edge, as is usual in wooden bridges. The keys also serve to throw the vertical component of the strain in the ties, directly upon the braces, without the intervention of the soft wooden chord.

The lower angle-block, or brace-bearing, is cast in a single piece, having a series of webs on the under side through which the pin passes.

The top laterals are of the pattern commonly used with the Howe truss, except that the bearing of the half struts is taken by small castings placed around the centre of the long strut, instead of being thrown directly upon the wood. The bottom laterals have cross struts and diagonal ties, each strut extending from the foot of a post to the opposite point on the chord-link of the other truss; the ties connect at one side with an eye-plate which fits over the chord-pin, and at the other with bent rods attached by nuts and a casting to

* These details are shown on Plate X., which contains also the general plans of the 248 foot span. Plate IX. contains a general elevation of the bridge, and some plans of the 176 foot span.

11

Engineered Irony

the chord-link ; each tie is in two parts, the adjustment being made with a sleeve nut.

The floor beams, of which there are two in each panel, are formed of two pieces of pine, each six by fifteen inches, placed side by side and trussed ; they are placed above the lower chord and rest upon cast-iron plates with raised centres, by which the weight is distributed equally upon the several links. Owing to the skew of the bridge, which is reduced to six feet three inches in the superstructure, the two floor beams which come in the same panel on one truss are divided on the other. The floor stringers, running lengthwise with the bridge, are seven in number, those under the rails being each formed of two pieces of seven by fourteen inch pine ; on these are laid two courses of one and a quarter inch matched flooring placed diagonally, the planks of the second course crossing those of the first, three layers of tarred paper, heavily coated with fresh roofing pitch, being placed between the two ; on these is laid a Nicholson pavement four inches thick, the whole being covered with sand and pitch in much the usual manner. The rail is of the street rail pattern, weighing 68 pounds to the yard, and made at the Palo Alto Rolling Mill, at Pottsville, Pa. ; it is laid on a longitudinal strip of oak resting upon the pine flooring. The floor is given a slight arch, and is drained into gutters on each side, which discharge through cast-iron scuppers, placed at such intervals as to avoid wetting the floor beams.

A foot-walk is placed upon the west side of the bridge, supported by brackets which are bolted to the floor beams ; the floor is made of two inch plank and a substantial wooden hand-rail is placed on the outside. The top chord is protected from the weather by a covering of pine boards, finished with a narrow overhanging cornice ornamented with brackets. The whole super-structure, including the iron parts, is painted with three heavy coats of a mixture of oil and crushed iron ore, manufactured by the Iron-Clad Paint Company, Cleveland, Ohio ; all cracks and weather checks in the timber having been stopped with putty, after putting on the first coat. The wooden keys and all joint-bearings were painted with the same composition before putting the truss together, and the closed covering is covered with a roofing paper made by coating thick brown paper with a coarse variety of this paint.

As the trusses are subject to a double deflection, the expansion of the lower chord under an increase of temperature operating in this way, as well as the strains produced by a load, they are built with a somewhat greater camber than is usually put in railroad bridges; the camber of the 248 foot span is 8½ inches, that of the 198 foot span 7 inches, and those of the 176 and 130 foot spans, respectively, 5½ and 4½ inches. These cambers, however, are materially in excess of any actual deflection.

In proportioning the trusses the central tie rods and the truss rods of the floor beams were allowed to bear a strain of 10,000 pounds to the square inch, each floor beam being supposed to take the greatest load which the drivers of a locomotive could possibly throw upon it, and no allowance being made for the stiffness of the timber under a transverse strain; the strain in the end ties and chord-links was limited to 12,000 pounds per square inch, but no allowance was made for the reduction of strain which the curvature of the upper chord would make in the end panels of the web. This practice of allowing a greater strain per square inch on those parts which are fully strained only under a maximum load, than on those which are liable to be strained to the full calculated amount by any heavy locomotive, is believed to have originated with Mr. Albert Fink, and is thought to be a more accurate method of proportioning than the common one, which makes no difference in strain per square inch on the different parts under a maximum load. The strain upon the timber of the top chord was limited to 800 pounds on the square inch, the braces were proportioned by the well-known formula of Hodgkinson, seven being the factor of safety adopted. The assumed moving load was 2,240 pounds per running foot for the four longest spans, and 2,800 pounds for the 130 foot span.

The trusses are anchored to the piers by long rods of one and a half inch round iron, which, extending from the top chord, pass over cast-iron struts projecting outwards from the coping, and are fastened by nut and screw through the eye of a three-inch pin set fifteen inches into the masonry. The trusses are further stiffened by corner braces extending from the end posts to a cross stretcher overhead, and the three longest spans have sets of similar braces placed at intervals through their length.

The amount of material in the several trusses, including floor beams and stringers, is as follows :—

LENGTH OF SPAN.	TIMBER.	WROUGHT-IRON.	CAST-IRON.
130	35,739 ft. B. M.	44,053 lbs.	27,137 lbs.
176	57,854 "	72,969 "	49,491 "
198	78,277 "	89,449 "	54,119 "
248	101,688 "	147,432 "	70,646 "

To this must be added 194,911 feet B. M. of pine lumber, 24,167 ft. of oak, 7,200 lbs. of wrought-iron, and 1,700 lbs. of cast-iron, used in the planking, pavement blocks, hand rail, vertical bracing, anchor rods and chord covering, on the fixed spans ; this additional amount includes the floor and footway of the 66 ft. span.

The method of manufacture by which the wrought-iron parts were prepared rendered them free from the common danger of defective welds. The chord links were made by upsetting the ends of flat bars of rolled iron till an increase of section somewhat in excess of that required was obtained, and then working down under the hammer and drilling the hole for the pin, leaving them absolutely free from welds. The only weld in the panel ties was that formed in joining the return end of the loop to the long bar ; this weld would at most be exposed to but half the strain upon the tie ; and even if the weld were absolutely worthless, the tie would have the full strength of a hook around the chord-pin. For these reasons, it was not considered desirable to test every piece of iron with a moderate strain of 20,000 to 25,000 lbs. to the inch, as is often done for similar works. Such a strain could at most be expected to reveal the defects of manufacture, which the methods here adopted precluded the possibility of ; while the effect of such a strain, by giving a set to the iron, and impairing its perfect elasticity, would be deleterious rather than otherwise. Samples of the iron were, however, taken and tested to breaking in an hydraulic tester under a slow and long-continued strain, with the following results :—

First Test.—Bar, $1\frac{1}{2}$ inches square, with welded loop at each end, length 5 feet between centres of loops. Four equal spaces of one foot each were laid out on the central part of the bar. No perceptible extension was noted with a strain of 10,000 lbs. per square inch, and but little with 20,000 lbs. With a total strain of 130,000 lbs., equal to 57,777 lbs. per square inch, the length had increased six inches, and the bar yielded by the opening of one of the welds; the four spaces, after the removal of the bar, measured respectively $12\frac{15}{16}$, $12\frac{11}{16}$, $12\frac{3}{4}$ and $13\frac{1}{2}$ inches; the last including a part of the broken weld.

Second Test.—Bar 2 inches square, 9 feet long, with loop on one end and nut on the other, extended by strain of 130,000 lbs. 1 inch, by 160,000 lbs. 3 inches, by 176,000 lbs. $4\frac{1}{2}$ inches, by 200,000 lbs. 8 inches, and by 221,000 lbs. 12 inches, when it broke about 18 inches from the nut, showing a fine fibre-like fracture, the strain being 55,250 lbs. to each original inch of section; but the reduced section at the break was only 2.85 sq. inches, making the strain somewhat over 77,000 lbs. to the square inch. Five equal spaces of one foot each, laid off on the bar before straining, measured after the break $13\frac{1}{8}$, $13\frac{7}{8}$, $13\frac{9}{16}$, $13\frac{7}{16}$ and $13\frac{1}{2}$ inches.

Third Test.—Bar $1\frac{3}{4}$ inches square, 38 feet long, *area* ~~and~~ 3.0625 sq. inches, under the following strain extended as given below :—

Total strain,	26,000 lbs.	8,500 lbs. to square inch.		Extension,	$\frac{1}{8}$	inch.
"	52,000 "	17,000 "	"	"	$1\frac{7}{64}$	"
"	78,000 "	25,500 "	"	"	$2\frac{5}{64}$	"

No perceptible set after strain of 28,000 lbs. on the square inch.

A number of tests were made at the same time of iron of the same quality manufacturing for the Dubuque bridge, with similar results. This iron is of the kind known as Kloman's mixture, manufactured at the Union Iron Mills, Pittsburg, the ties and truss-rods being made of double-rolled, refined iron. The bar broken in the second test was afterwards cut up, and three small pieces were turned out of it, each having a reduced central diameter. These were taken to the Fort Pitt Foundery, and tested in the lever machine belonging to the U. S. Government; two of these tore out of the clutches before breaking, when the strains per square inch were respectively 62,760 lbs. and 63,134 lbs.

The last specimen broke under a strain of 84,032 lbs. per square inch, and showed a beautiful fracture entirely fibrous.

The cast-iron used in the details was a gray iron formed of a mixture of pig, generally adopted by the Keystone Bridge Company.

Specimens were tested by suspending a weight upon a bar two inches by one, and placed upon supports four feet apart. The specifications required that this breaking weight should not be less than 2,100 pounds, and in all of the tests it was found to be much in excess of this amount.

The shore span is a riveted trellis girder of wrought-iron 71 feet long and 8 feet deep. The chords are of $\top$ section, composed of two vertical plates, one horizontal plate and two angle pieces; in the bottom chord the horizontal plate does not reach to the ends of the span, and the other parts are continuous for the whole length. The braces are each formed of two pieces of $\top$ iron placed back to back, and enclosing the ties, which are single bars of flat iron; both ties and braces are riveted between the vertical chord plates. The laterals are of wrought-iron, and the trusses are stiffened by short braces of $\top$ iron connecting the floor with the web. The end posts are enclosed in light ornamental castings. The floor beams are of pine, six inches by eighteen, without trussing, placed two feet between centres; on this is laid a floor similar to that on the other fixed spans. The amount of material in this span, exclusive of pavement and hand rail, is as follows: Lumber, 7,684 feet B. M.; wrought-iron, 32,165 pounds; cast-iron, 4,328 pounds.

The draw measures 361 feet and 3 inches over all; it is a Pratt truss of similar plan to the large draws erected by the Keystone Bridge Company at Cleveland, Dubuque, and other points. The skew is taken out of the truss by making the end panels of unequal lengths, the difference being 5 feet 6 inches. The upper and lower chords are of like pattern, formed of two I beams and two channel bars eight inches deep, placed side by side and united by a plate riveted to their upper flanges. The posts are of wrought-iron, of the Linville pattern. The ties are round, with both ends upset for screws; the main ties are in pairs, and the counters single, passing through the posts. The washer plates upon which bear the nuts of the ties are of cast-iron, except the top centre, which is forged. The floor beams are ten-inch rolled I beams, and rest on the top of the lower chord. The floor is of two-inch oak plank laid on the oak track stringers, and pine floor joists. There is no separate footway on the draw. The turn-table is formed of an external drum thirty feet in diameter, and a central shell of cast-iron, hung by ten bolts on one of Sellers' patent pivots ; the drum and shell are connected by a pair of plate girders under the centre posts, and a set of radial rods. The bolts are adjusted so as to throw almost the entire weight on the centre, the drum serving only as a guide and balancer. The draw is easily opened by four men, with levers attached to two pinions on the drum, in two minutes, but as a precaution against wind and other dangers, it is to be fitted with a steam-engine. The latch is worked from the centre by a hand-lever ; a bearing is secured by wedges which are driven under the four end-posts, the four being worked by a single central lever. The amount of material in the draw, including both trusses and turn-table, is as follows : Timber (in floor), 26,025 feet B. M. ; wrought-iron, 495,575 pounds ; cast-iron, 122,041 pounds.*

In proportioning the draw, it was supposed to carry the whole dead load on the central bearing when swung, and each arm was supposed to carry its share of the dead load, and a moving load of one ton to the foot when closed, no allowance being made for the continuity of the chords. Though this has been the method by which most of the large iron draws lately built have been proportioned, the engineers were convinced that it is a method of computation which gives very erroneous results, showing the central strains, especially in the web, to be much less than they really are, with corresponding excesses in other parts ; a set of calculations believed to be based on a more correct hypothesis will be found in a subsequent chapter. The distribution of strain is regulated by the proportion of the total weight thrown upon the end piers, and is there-fore largely dependent on the form of latch used. The wedges under the end posts have but a small lifting power, as is fully proved by the action of the draw under a passing load, a heavy freight train, covering one arm only, causing the further end to rise from its bearings $\frac{7}{8}$ of an inch. A set of hydraulic jacks are to be substituted for the wedge plates, the jacks being placed within the hollow end-posts and worked from the turn-table by pumps driven by the steam-

---

* The Plans of the Draw are given on Plate XI.

engine; it is thought that under this arrangement a sufficient lifting power can be obtained to make the proportioning of the draw sufficiently correct to prevent distortion.

The 130 and 176 foot spans, as well as the little iron span, were raised in the spring of 1868. The 130 foot span was the first erected, the trestle used in building Pier No. 1 being made available for one side of the false-works. The two other spans were over dry land at the time of their erection, and ordinary false-works, resting on the ground, were used for their raising. The remainder of the superstructure was not raised till the spring of the following year, when the first span raised was that between Piers 5 and 6, while the sand bar continued dry. The greatest difficulties occurred in the case of the span between Piers 3 and 4, where the strength of the current and depth of the water, especially near Pier No. 3, would have carried away any common false-works in a very few hours. The distance between the caisson around Pier No. 3 and the false-works at No. 4 was divided into four nearly equal spaces. Between the first and second of these spaces, a cluster of eight piles in two rows eight feet apart was driven in thirty-five feet of water, the piles being kept from washing out by guying them with lines as fast as driven; a crib of round timber was then built, enclosing the piles, which, on being sunk by filling it with stones, should at once retard the wash and bind the piles together. A precisely similar arrangement was adopted between the second and third spaces. This work was begun on the 10th of March; on the 14th the weather became very cold, and the ice began to run in large quantities; the numerous obstructions of the false-works impeded the flow of ice, and in the forenoon of the 16th it jammed at the bridge site and the river became closed. The weather had already begun to moderate, and in the afternoon of the same day the ice moved out; it was very weak, but the cakes had packed together, forming large thick fields, which, however, were too soft to bear the weight of a man. The sixteen piles of the two clusters had been driven, a crib had been built about the first cluster, though not sunk, and carpenters were at work upon the second crib, when the ice began to move across the whole width of the river at once; it tore out all of the sixteen piles, taking the cribs with them, and carried along with it the pile-driver, barges and men. The boats moved but slowly, being frequently retarded by ice jams, and while still opposite the town they were overtaken by the steamboat and secured, having suffered no material damage ; but no trace of piles or crib-work remained, and two months later one of the cribs was observed forty miles down the river, with a pile still remaining in it. This gorge was accompanied by a considerable scour, the water at the site of the second set of piles having been deepened about twelve feet.

The piles were at once replaced, and the cribs built, sunk and protected by additional riprap ; the piles were then capped and surmounted by trestle piers, which were planked on the sides and provided with timber starlings, eight or ten feet high, as a protection against drift. Between the third and fourth spaces, where the depth of sand was much greater, a single row of five piles was driven, which were braced to the false-works of Pier No. 4, riprapped and surmounted by a trestle bent. Other bents were raised on the caisson surrounding Pier No. 3, and on the false-works at No. 4, and a false pier of timber was erected immediately south of Pier No. 4, the masonry being still unfinished, one side of which rested on the upper section of the caisson, and the other side on the false-works. Eight trussed girders, made of track stringers and rods which had been used at No. 4, were built upon the shore, and raised by a floating derrick into position on the trestles, one being placed under each bridge chord ; on these were laid the cross-timbers and other staging required. These false-works proved amply stiff, and when removed after the erection of the span, it was found easier to break the piles off immediately above the cribs than to withdraw them. The false-works between Piers 4 and 5 were built at the same time, resting on piles, and a light track was laid from the north end of the bridge, nearly to Pier No. 3.*

The 198 ft. span was raised as soon as these false-works were ready. As Pier No. 4 was still incomplete, the last panel was not put in, but a bearing was taken on the false pier, one panel short of the end of the truss, the top chord stood projecting over, the links of the bottom chord being left to hang down. On the completion of the pier the last panel was added, and the bearing was transferred to the masonry. The erection of the long span between Piers 4 and 5 followed, completing the number of fixed spans.

---

*These false-works are shown on Plate VII.

The draw span was raised on false-works extending from the pivot pier to the upper and lower rests. As the small amount of sand above the rock precluded the driving of piles, these works were built on cribs, two of which, loaded with stone, were placed between the pier and each rest. These cribs were originally intended to serve as the foundation of a permanent draw protection; they were built in the winter of 1868-9; were made thirty feet square, and divided by four cross-walls into nine compartments. The deadening effect of the upper rest and pier on the current, had so checked the scour that the cribs did not settle to the rock, and as their bearing was not thought to be firm enough to carry a permanent structure, they were built up above ordinary high-water, and a wooden truss, strong enough to sustain itself if the cribs settled, and which should serve as false-works for raising the draw, was built upon them.*

As soon as Pier No. 2 had been completed, the pivot was placed upon it and the turn-table put together; the chords were then spread out and riveted, and the bridge trusses made self-sustaining at the earliest possible moment, the whole structure being raised in about six weeks. The cribs settled slightly under the weight of iron, but not enough to give trouble, the subsidence being remedied by additional blocking. Since then the upper cribs have not settled materially, and are probably on their permanent bearing; but the night after the weight of the truss had been taken off the false-works, a rise in the river scoured around the two lower cribs, causing them to settle away from the truss; under the continued scour of the summer flood they continued to settle, tilting from side to side, and finally, when the flood was at its height, they tipped over and rolled away; the false-work truss remains standing, and no harm was done to the works.

On the occasion of the public opening on the 3d of July, the bridge was tested in the presence of a number of engineers invited to examine it, with the following results :—

176 FOOT SPAN.

Load at North quarter of Span ....46 Tons.
North quarter Deflection ....$1\frac{1}{4}$ Inch.
Centre   "   ....$1\frac{9}{16}$ "
Load at Centre and North quarter ....92 Tons.

---

* The temporary Draw Protection is shown on Plate VII.

North quarter Deflection..$2\frac{4}{4}$ Inch.
Centre   "   ....$3\frac{1}{32}$ "
South quarter   "   ....$2\frac{8}{32}$ "
Load at Centre and both quarters ....112 Tons.
North quarter Deflection..$2\frac{6}{32}$ Inch.
Centre   "   ....$3\frac{4}{32}$ "
South quarter   "   ....$3\frac{3}{32}$ "
Full load ....170 Tons.
North quarter Deflection ....$3\frac{7}{32}$ Inch.
Centre   "   ....$3\frac{7}{32}$ "
South quarter   "   ....$3\frac{7}{32}$ "
Permanent Set ....$\frac{1}{16}$ "
Elongation of Bottom Chord ....$\frac{1}{4}$ "

198 FOOT SPAN (Piers 5 to 6).

Loaded with four Locomotives ....187 Tons.
North quarter Deflection..$4\frac{2}{4}$ Inch.
Centre   "   ....$5\frac{4}{32}$ "
South quarter   "   ....$4\frac{1}{32}$ "
Permanent Set...$1\frac{1}{32}$ "
Elongation of Bottom Chord ....$1\frac{3}{32}$ "

248 FOOT SPAN.

Fully loaded ....233 Tons.
North quarter Deflection..$5\frac{1}{32}$ Inch.
Centre   "   ....$6\frac{4}{32}$ "
South quarter   "   ....$4\frac{8}{32}$ "
Permanent Set ....$\frac{6}{32}$ "
Elongation of Bottom Chord ....$\frac{3}{8}$ "

198 FOOT SPAN (Piers 3 to 4).

Fully loaded ....187 Tons.
Centre Deflection ....$4\frac{3}{32}$ Inch.
No Permanent Set.

DRAW SPAN.

North Arm loaded ....170 Tons.
North quarter Deflection..$2\frac{3}{32}$ Inch.
Centre   "   ....$3\frac{4}{32}$ "
South quarter   "   ....$1\frac{5}{32}$ "
Centre of South Arm rose ....$\frac{19}{32}$ "

Both Arms loaded ....313 Tons.
    Centre Deflection, North Arm ....$\frac{19}{32}$ Inch.
    Permanent Set,   "   " ....$\frac{1}{8}$ "
    North quarter Deflection, South Arm......................$\frac{1}{3}\frac{5}{2}$ "
    Centre   "   "   " ....$\frac{1}{3}\frac{8}{2}$ "
    South quarter   "   "   " ....$\frac{9}{32}$ "
    Permanent Set..$\frac{3}{8}$ "

It is to be noted that these tests were made before the bridge had been screwed up under a load, and while the bearings were not perfectly close ; they consequently show greater deflection than a subsequent testing would indicate.

CHAPTER VI.

OUTFIT.

THE isolated position of Kansas City, its distance from manufacturing and commercial centres, and the unsettled character of the neighboring country, were most felt at the beginning of the work in the collection of a suitable outfit. The character of the work was also such that the full number of tools needed was only learned as the works advanced, and the greater part of the equipment grew up with the progress of the several foundations. Nearly all the tools, including the boats used for transporting material, had to be built for the purpose, much of the time during the spring and summer of 1867 being occupied in this preparation. The derricks were built of native lumber, which was but poorly suited to this object, though it was made use of as far as could well be done ; the masts and booms were made of cotton-wood, and though of a cheap pattern, the derricks did good service. A machine-shop was fitted up on the bank of the river, two-thirds of a mile above the bridge, where all the smaller iron-work needed, and the lighter pieces of timber-work, were prepared ; this shop was fitted out with a Daniels' planing-machine, circular-saw, lathe, screw-cutter, drill, and such other implements as were required for four blacksmith fires. Though containing a number of tools not usually wanted for bridge works, the outfit was as small as was deemed practicable, and could more properly be charged with inadequacy than with extravagance.

The boats provided were as follows :

A side-wheel high-pressure steamboat, 135 feet long, of the kind in common use on Western rivers ; this boat, "The Gipsey," was built for an Ohio river packet, on which river she was purchased.

Four flat boats, each 53 feet long, 18 feet wide, and 3 feet deep.

Engineered Irony

One smaller flat boat, with hole through the centre for boring, and used also for stone.

Two small barges, about 50 feet long, purchased at Kansas City, and rebuilt for use on the bridge.

Three flat boats with square bows, two of them 20 by 53 feet, and the other 22 by 64 feet. Two of them were fitted out as floating derricks,* and the third was used for pile-driving.

One small scow, housed, and used as a diving boat.

A large yawl, with a crew of seven men, and a skiff manned by two men, were also kept on the river; a few other skiffs were generally in use on various parts of the works, and two small flats, which could easily be towed by the yawl, were used as sand boats. One of the large flats sunk in the spring of 1868 when heavily loaded with stone, and was lost; the other six were rebuilt after the completion of the bridge; their decks were raised, and they were converted into pontoons, to be anchored in a line above Pier No. 1, where they should serve as a protection for steamboats passing down through the draw.†

The principal items in the remaining equipment were :—

Four portable steam-engines, of 8, 12, 15 and 25 horse-power respectively; one of these was employed to drive the machinery in the shop.

One nine-inch Alden centrifugal pump.

One No. 4 Andrews centrifugal force-pump, four inch discharge, and six inch supply pipes, with flexible hose.

One six-inch steam siphon pump.

One donkey pump, used for jetting.

One air-pump, diving-dress and outfit complete.

Four large dredges, with attachments.

One small dredge.

Three pile-drivers, with 2,200 pound hammers.

Eight land derricks, with rigging complete, and horse-powers.

One sawing-machine, for cutting off piles under water.

---

* Plate VIII.

† The Pontoon Protection is shown on Plate II; the dotted lines on the plan indicate the position of the floats at high water.

Three triangular beton boxes.

Two square beton boxes.

Two steam crabs, one single and one double.

Six hand crabs.

To these should be added the belts, pulleys, and shafting used in driving the machinery at the several piers, the long suspension screws referred to in the chapter on Foundations, with nuts and wrenches ; the gas-pipe and flexible hose, used for both steam and water ; sand-pans and water-kettles used in mixing mortar in cold weather ; two or three small portable forges ; a good supply of blocks and lines, and the proper complement of smaller tools, which it is needless to mention. A tremmie was constructed for laying beton under water, but never used. The more important tools which were designed originally for this work, have been described in the preceding pages.

CHAPTER VII.

CALCULATED STRENGTH.

The many plans prepared during the construction of the Kansas City Bridge involved a proportionate amount of mathematical calculation ; much of this was of a simple and elementary character, devoid of general interest ; but no account of the work would be complete which did not embrace a review of the stresses in the foundation works, the pressure upon the several foundations, and the strains in the superstructure, these being the points in which the computations were carried into the greatest detail, and which have the most important bearing on the general structure.

The foundation works embrace both the caissons, which were exposed to the pressure of the sand and water, and the upper works, which carried the suspended weight. The strains in the latter were of a simple character, and need not be enumerated here ; those in the former arose from the pressure of the water, due both to the current and the depth, and the pressure of the sand, including also the friction caused by this pressure on the sides of the descending caisson. The effects of the current was computed, but, though important in determining the strength of the cables used in anchoring the water-deadener and placing the round caisson, it was too slight to influence the general results elsewhere.

WATER PRESSURE.

The greatest water pressure occurred when the caisson for Pier No. 1 was pumped out. The surface of the water was then 101.4, about a foot and a half above the ordinary low-water stage, and four feet and a half above the extreme low-water ; the elevation of the rock was about 84, so that the pressure corresponded to a depth of very nearly 17.5 feet ; this made the pressure on each horizontal foot of caisson 9,570 pounds, and the total pressure, estimating the perimeter at 155 feet, 1,483,250 pounds, or a little less than 750 tons. The form of the caisson was such that the starlings braced themselves, and the only pressure which had to be carried by interior braces was that on the opposite long walls, the total strain on the braces being equal to the pressure on 55 feet, the length of one of these sides, or 483,300 pounds. This would have been carried by twenty-five braces, each eight inches square, with a strain scarcely exceeding 300 pounds on the square inch; but to avoid all possibility of accident, nearly double this strength of bracing was used. No other caisson was pumped out to nearly this depth; the round tub used at Pier No. 2, from its circular form, withstood the strains upon it without the aid of interior bracing.

SAND PRESSURE AND FRICTION ON SIDES OF CAISSONS.

The pressure of the sand was considered the same as the thrust of a bank of earth, the particles of which have no mutual cohesion and computed by the formula :—

$$P = \frac{wh^2}{2} \tan^2 \frac{a}{2} * \qquad (a.)$$

in which P denotes the total pressure on each horizontal foot; w, the weight of a cubic foot of the earth or other material; h, the height of the bank in feet, and a, the angle which the natural slope of the material makes with a vertical line, being the complement of the angle of repose and determined by the relation :—

Cot. a = coefficient of friction of material on itself.

The application of this formula becomes somewhat complicated when the earth or sand is submerged. The action of the water is threefold : 1st, it gives buoyancy to the mass, thereby diminishing the weight; 2d, by acting as a lubricator on the surfaces in contact, it reduces the friction and increases the value of a; 3d, the pressure due to its weight is added to the thrust of the bank. The two first of these are simple and easily provided for by making the proper changes in the values of w and a; the latter is of a more complicated nature, dependent largely on the character of the material. If the bank

---

* This formula is taken from Claudel, Aide Memoire, etc. 7ieme Edition, p. 1252. It is also found in a slightly modified form in Rankine's Civil Engineering, 4th Edition p. 322 (11.)

is formed of loose stones or coarse gravel, the pressures of sand and water remain distinct, each substance transmitting its own pressure, and the gravel alone producing friction ; if the material is a water-tight clay, the weight of the water above is equivalent to that of any other load, increasing definitely the pressure of the clay and causing additional friction ;* if, however, as is commonly the case, the material be something between these extremes, a fine sand, a silt, or mixture of the two, the water neither penetrates the whole with perfect freedom nor remains as a weight on the top of a substance which it does not enter ; its action is therefore dependent on capillary attraction and matters which cannot be measured precisely ; and while the total pressure would not differ materially from that in either of the two preceding cases, the portion of that pressure which is transmitted by the sand, and which alone produces friction, would be somewhat greater than in the case of the gravel and less than with the clay. This could be better guarded against by an empirical allowance than measured by exact computation ; in estimating friction, accordingly, the calculations were made by the formula given above, but the value given to w was the full weight of the saturated sand, and not its submerged weight alone ; this, undoubtedly, gave excessive results, but as no allowance was made for the portion of the pressure of the superposed water transmitted by the sand, this discrepancy was a little less than might at first be supposed.

The coefficient of friction of wet Kansas City sand upon itself was ascertained by experiment to be about .8 ; the least observed was .725, which corresponds very nearly to $a = 54°$; this, substituted in the above formula, gives

$$P = 0.1298 \; h^2 \; w. \tag{b.}$$

Substituting for w the immersed weight of the heaviest sand weighed, or 69.5 pounds :—

$$P = 9.02 \; h^2. \tag{c.}$$

And if w be made the full weight of a cubic foot of such sand saturated,

$$P = 17.13 \; h^2. \tag{d.}$$

---

* In this case the actual and not the submerged weight of the clay must be used in computation ; but the cohesion of the particles of clay is so great that this formula would give very excessive results.

THE KANSAS CITY BRIDGE.

Experiments indicated the coefficient of friction of dressed oak on sand to be .475, but in calculations it was generally assumed to be .5, which, substituted in equation (c.), gives for the friction corresponding to each horizontal foot of caisson ;—calling the friction F

$$F = 4.51\,h^2; \tag{e.}$$

or, substituted in equation (d),

$$F = 8.56\,h^2. \tag{f.}$$

The average weight of saturated sand, however, did not exceed 125 pounds to the cubic foot, and the coefficients of friction adopted have been slightly excessive ; the decimals may therefore be omitted, and the formula reduced to the convenient form :—

$$F = 8h^2. \tag{g.}$$

The average friction in pounds on each superficial foot of caisson in contact with the sand may therefore be considered as eight times the average depth in feet of the cutting edge below the surface of the sand. This formula of course varies with the material, and in its present form is applicable only to the Missouri River.

The sand pressure on the caisson at Pier No. 5, when sunk twenty feet into the sand, that being the depth of sand immediately around it when the sinking was completed, computed by formula (c.), which would properly be used in this case, as the external water pressure, whether through sand or water, was balanced by an equal internal water pressure, was 3,608 pounds on each horizontal foot, or, estimating the perimeter as 155 feet, 559,240 pounds on the entire caisson ; this was less than two-fifths of the water pressure on the caisson used at Pier No. 1, and was easily carried by internal braces. The sand pressure at No. 3 was never so great as this. In proportioning the inverted caisson for Pier No. 4, the timber-work was made strong enough to withstand the thrust of the sand, without the assistance of the beton. The formula by which the pressure should be computed in this case is :—

$$P = 9.02\,(h^2 - h'^2). \tag{h.}$$

in which h denotes the total depth of sand, and h' the depth above the top of the inverted caisson. Assuming $h = 40$ and $h' = 28$, this equation gives for the pressure on each foot of perimeter 7,360 pounds, a weight which a timber wall at least fourteen inches thick and eleven feet wide would easily carry over the distances between the three cross-walls.

The planking of the caisson used for Piers 3 and 5 was not dressed, and the roughness of the timber increased the friction about one-quarter, changing formula (*g.*) to

$$F = 10h^2. \tag{$i.$}$$

The available weight was in each of these cases barely enough to overcome this friction, which accounts for the slow progress of the sinking and the interruptions caused by sand-slides. The greatest available weight of the caisson at Pier No. 3 was about 700,000 lbs., which is equivalent to the friction produced by 21.25 feet of sand, the perimeter of the caisson being 155 feet; though this was greater than the actual average depth of sand, the excess was too small for advantageous work. At Pier No. 5, on the 2d of July, 1868, the effective weight for each foot of perimeter of the caisson was 2183.5 pounds; the average depth of the surrounding sand was then 15.5, corresponding to a friction of 2402.5 pounds per horizontal foot; this deficiency was remedied by piling sand above the caisson, but the weight was always too small for good results. It may be noted, that a caisson whose weight is barely greater than that of the water it displaces, may be sunk by long-continued dredging; the amount of sand excavated will be many times the capacity of the caisson, but, as the external sand slides down and passes under the edge, it will slowly carry down the caisson.

The relation between weight and friction at Pier No. 4 is most plainly shown by the tables in Appendix E; the friction per square foot of rubbed surface, computed by formula (*g.*), is added to these tables for convenience in showing this relation. The advantage of having a sufficient excess of weight to cause the cutting edge to penetrate well into the sand, cannot be overestimated : it aids in feeding the excavators, reduces the amount of excavation, and precludes sand-slides.

PRESSURE ON FOUNDATIONS.

The pressure upon the foundations of the seven piers is given below. In these computations the masonry is assumed to weigh 155 pounds per cubic foot, and the beton 135 pounds; these weights are probably slightly in excess of the actual weights. In estimating the weight of the superstructure carried, Piers 1 and 3 are each supposed to carry one half the weight of one arm of the draw, with the same length of moving load, this result being attainable, with a lifting latch of sufficient power; Pier No. 2 is supposed to carry the entire weight of the draw and turn-table, with 290.5 feet of moving load, those being the weights carried by it under the present arrangement of wedge plates. In Piers 1, 2 and 3, the base of the foundation is assumed to be the whole size of the caisson inside of the frame, but not reaching within fifteen inches of the outside of the planking; in Pier No. 4, where the lower caisson forms not only the covering, but an integral part of the pier. the foundation is assumed to be the full size of the caisson, and the timber and iron are computed as part of the weight carried. The moving load is assumed to be 2,240 pounds to the foot, excepting in the case of Pier No. 7, where, as the length of track carried is but 100 feet, the load per foot is estimated at 2,800 pounds; the weights of the trusses are the same as those used in proportioning the superstructure :—

PIER No. 1.

Masonry, 1,234 c. yards	5,164,290 lbs.
Truss	131,500 "
Draw, 90 feet	172,800 "
Moving load, 157.5 feet	352,800 "
Total	5,821,390 "

Area of base, 986 sq. feet; pressure per sq. foot, 5,904.05 lbs.; pressure per sq. inch, 41 lbs.

PIER No. 2.

Masonry, 1,199 c. yards	5,017,815 lbs.
Beton, 767 c. yards	2,795,715 "
Draw	735,000 "
Moving load, 290.5 feet	650,720 "
Total	9,199,250 "

Area of base, 1,104.46 sq. ft.; pressure per sq. ft., 8329.18 lbs.; pressure per sq. inch, 57.84 lbs.

Engineered Irony

PIER No. 3.

Masonry, 873 c. yards	3,653,505 lbs.
Beton, 833 c. yards	3,036,285 "
Truss	245,000 "
Draw, 90 feet	172,800 "
Moving load, 181.5 feet	406,560 "
Total	7,514,150 "

Area of base, 1,087.12 sq. ft. ; pressure per sq. ft., 6,911.92 lbs. ; pressure per sq. inch, 48 lbs.

PIER No. 4.

Inverted caisson	291,625 lbs.
Section 2, "	95,810 "
" 3, "	95,150 "
Masonry, 1,109 c. yards	4,631,165 "
Beton, 1,169 c. yards	4,261,005 "
Trusses	572,180 "
Moving load, 225 feet	504,000 "
Total	10,450,935 "

Area of base, 1,321.87 sq. ft. ; pressure per sq. ft., 7,906.15 lbs. ; pressure per sq. inch, 54.9 lbs.

PIER No. 5.

Masonry, 820 c. yards	3,421,700 lbs.
Timber footing	100,000 "
Trusses	572,180 "
Moving load, 225 feet	504,000 "
Total	4,597,880 "

144 piles ; weight per pile, 31,929.17 lbs.

PIER No. 6.

Masonry 656 c. yards	2,776,016 lbs.
Timber footing	100,000 "
Trusses	452,900 "
Moving load, 188 feet	421,120 "
Total	3,750,036 "

90 piles ; weight per pile, 41,667.07 lbs.

PIER No. 7.

Masonry, 434 c. yards	1,816,290 lbs.
Timber footing	100,000 "
Truss	207,900 "
Moving load, 100 feet	280,000 "
Total	2,404,190 "

73 piles; weight per pile, 32,934.11 lbs.

RESISTANCE AGAINST ICE.

The shock imparted to a pier by the impact of ice, depends upon the size of the cake, the speed of the current, and the time occupied in bringing it to rest; if a small cake could be stopped instantaneously, it would impart a sufficient shock to move the largest pier a distance proportional to their relative size, but as the cake is always more or less shattered by the blow, it moves a greater or less distance after it strikes, and the time which it continues to move, or, measuring this motion by distance, the distance which the pier ploughs into the field of ice reduces and determines the force of the impact. The greatest shock which a pier will be called upon to sustain might be calculated from these conditions, but can be more readily estimated by measuring the force of a blow which will crush the ice along the whole width of a pier. Sound ice begins to yield under a pressure of 200 pounds on the square inch, and is crushed to atoms when this pressure is increased to 450 pounds, as was ascertained by experiments made on these works. When the ice breaks up at Kansas City it is seldom more than one foot thick, and the upper part is always so soft and rotten that one foot may be estimated as the maximum thickness of solid ice. Piers 1 and 3 finish eleven feet thick at an elevation of 112, fifteen feet above extreme low water; this is probably some feet higher than the ice ever moves; the amount of masonry above this elevation is 405 yards, which, with the portion of the truss carried by Pier No. 1, weighs, estimating the masonry at 4,000 lbs. to the cubic yard, 1,751.500 lbs. The crushing shock on eleven square feet at 450 lbs. to the square inch is 712,800 lbs., and the ratio of this shock to the weight that must be moved to disturb the pier is .4175, or less than the coefficient of friction of stone on stone; the piers are therefore sufficiently heavy to withstand the blows from floating ice, without relying upon the adhesion of the cement, or depending upon the retreating and pointed form of the ice-breaker for any further assistance than to keep the pier clear of jams.

THRUST OF SAND BARS.

Another disturbing cause to which the piers are exposed is the thrust of a sand bar on one side, while the other side is washed clean by the scour. The greatest exposure of this kind would occur at Pier No. 4; if the sand-bank on the north side of this pier were forty feet high, and the rock was swept clean on the south side, the thrust of sand to be sustained on each horizontal foot, estimated by formula (c.), would be 14,432 pounds, and the thrust upon the whole pier, 70 feet long, would be 1,010,240 pounds. The moment of this thrust, tending to overturn the pier, will be equal to this pressure multiplied by the height of the equivalent centre of application above the base of the pier, or one-third the height of the sand-bank, 13.33 feet; which gives for the over-turning moment 13,469,833 pounds.

The weight of the pier, allowing for the buoyancy due to the immersion in an extreme flood, is not far from 6,000,000 pounds. The moment of weight tending to resist overturning is equal to this amount multiplied by one-half the breadth of the base of the pier, or 11.25 feet, giving for this moment of resist-ance 67,500,000 pounds, or about five times the greatest overturning moment due to the thrust of the sand bar.[*]

STRENGTH OF FIXED SPANS OF SUPERSTRUCTURE.

The strains on the fixed spans of the superstructure were computed by the simple method usually employed in calculating the strains on the different mem-bers of an isolated span. The strains upon the panel ties and braces in all the spans were estimated as if both chords were horizontal; the results thus obtained were not strictly correct in the cases of the three longest spans, but

---

[*] No allowance has been made in this calculation for the bond formed by the beton with the bed rock, whereas, even if the adhesion of the mortar be considered as nothing, by excluding the water from the base of the pier, it makes the entire weight of the pier, without deduction for immersion, available to prevent its overthrow.

the errors were on the side of safety, and the strains thus calculated in excess of those which actually occur. This will appear from the accompanying figures.

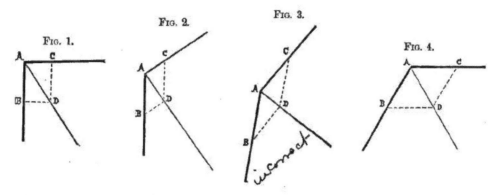

Figs. 1 and 2 represent the upper parts of the end panels of single triangular trusses in which the end post is vertical, and Figs. 3 and 4 the same parts of similar trusses in which the end post is inclined. In Figs. 2 and 3 the upper chord is inclined as in the end panels of a truss in which the upper chord is arched, and in Figs. 1 and 4 it is horizontal. As the end posts carry the full weight transferred to the masonry, the weight upon them will be the same in every case ; representing the strain upon these parts by AB, and completing the parallellograms of forces A B D C, it is evident, by an inspection of the figures, that the diagonal AD, representing the weight carried by the panel tie, is less in Figs. 2 and 3 than in Figs. 1 and 4. The same will be true in every panel where the arch of the upper chord gives it a sufficient inclination to have an appreciable effect, and as the double triangular truss is simply a combination of two single triangular trusses, in one of which, according to the arrangement adopted in this bridge, the end post is vertical and in the other inclined, the strains in the web will be reduced by the action of the arch in the same manner.

The maximum strains upon the different parts of the fixed spans under the action of a moving load, computed as above stated, are given in Appendix F.*

---

* Skeleton Diagrams of the 130, 176, 198 and 248 ft. spans are given on Plate XII. The figures denote the strains in tons of 2,000 lbs., the compressive strains being marked +, and the tensile —.

DEFLECTIONS OF THE 248 FT. SPAN.

The 248 foot span measures 246 feet between the centres of end connections, the height between centres 22 feet, and the versed sine of the arch of the upper chord is 10 feet. A full load of 2,240 pounds to the foot, or 250 gross tons for the entire span between centres of piers, throws upon the timber of the top chord a compressive strain of 325 pounds per square inch of section, and on the wrought-iron of the lower chord a tensile strain of 5,200 pounds per square inch of section; these strains being in addition to those caused by the dead load. The compressive strain of 325 pounds per square inch on the pine timber shortens the upper chord .000226 of its length, and the tensile strain of 5,200 pounds per square inch on the wrought-iron links lengthens the lower chord .00018 of its length. The deflection caused by the load depresses the centre of the arch, thereby increasing the distance between the end connections; denoting the deflection by d and this increase of distance by e, it will be determined by the equation

$$e = d\frac{2 \times 10}{123} + \tfrac{1}{4}d\frac{4 \times 2.5}{61.5} + \tfrac{1}{16}d\frac{8 \times .5}{30.75} + \ldots = \frac{26\,d}{123}$$

The radius of the curve assumed under this deflection by a horizontal line connecting the ends of the span is given by the equation

$$r = \frac{h}{i - i'}$$

in which h denotes the height of the truss at the ends, i the relative increase of length in the bottom chord, and i' the relative increase of distance between the end connections of the top chord; in this case

$$h = 22 \qquad i = .00018 \qquad i' = -.000226 + \frac{e}{246}$$

The deflection is given by the equation

$$d = r - \sqrt{r^2 - 123^2}$$

The solution of these several equations gives

$$e = .0222 \qquad\qquad i' = -.000136$$
$$r = 69620 \qquad\qquad d = .109 = 1\tfrac{5}{16} \text{ inches.}$$

To obtain the actual deflection, that due to the strains in the braces and panel ties must be added to the deflection thus obtained. The heaviest strain caused by the moving load alone upon the braces is 240 pounds per square inch, and upon the ties 5,000 pounds per square inch, shortening each brace .000165 of its length, and lengthening each tie .000172 of its length ; making the deflection due to each triangle .000337 of the height of that triangle ; but as the central ties and braces, which are the longest, are but slightly strained by a full load, the average deflection due to all the triangles will not exceed .00025 of their height. The aggregate height of the five triangles on each side of the centre, in the system whose tie-rods meet at the centre, is 140 feet, making the deflection due to the strains in the web .035 feet, or $\frac{1}{3}\frac{1}{2}$ of an inch ; this, added to the deflection due to the chord strains, gives for the total deflection .144 feet, or $1\frac{2}{3}\frac{3}{2}$ inches.*

The unequal expansion of the wood and iron in the structure, under a change in temperature, causes the centre of the truss to rise and fall in a manner similar to the action produced by a passing train. The range of temperature at Kansas City may be assumed at 120° Fahrenheit ; in exceptional seasons it may exceed this, but only rarely. The coefficient of expansion of pine wood, for one degree of Fahrenheit, is .00000227, and that of wrought-iron is .00000698 ; then give for the values of i and i' the equation given above :—

$$i = .00000698 \times 120 = .000837$$
$$i' = .00000227 \times 120 - \frac{e}{243} = .000272 + \frac{e}{246}$$

The values of the other known quantities will be the same as in the preceding calculation. Solving the equations as before, we have for the effects of an increase of temperature of 120° Fahrenheit :—

$$e = .0317 \qquad\qquad i' = .000401$$
$$r = 50459 \qquad\qquad d = .150 = 1\frac{1}{1}\frac{3}{6}\text{ inches.}$$

The same increase of temperature acting upon the web causes a deflection in

---

* To secure strictly accurate results, the deflection caused in the top chord by the web strains should be calculated separately, and the general deflections corrected to correspond with the spreading of the arch due to them ; this difference, however, is but slight, and need not ordinarily be considered.

each triangle equal to the difference between the expansions of the tie and brace, and causing a deflection at the centre equal to

$$140 \times [.000837 - .000272] = .079 = \tfrac{15}{16} \text{ inches.}$$

The total deflection caused by an increase of temperature of 120° will therefore be .229 feet, or $2\tfrac{3}{4}$ inches.*

The extreme vertical variation of the centre of the span, from its position when unloaded at the lowest temperature, and when fully loaded at the highest temperature, will not exceed .373 feet, or $4\tfrac{15}{32}$ inches.

The deflections of the other spans may be calculated in the same manner, but will be less than the above. The calculations for the 176 and 130 ft. spans are much simplified by the absence of curvature in the upper chords.

STRAINS IN THE DRAW.

The weight of the iron and timber in the draw is 735,000 pounds; the turn-table, entirely of iron, weighs 52,000 pounds, leaving for the weight of the trusses and floor 683,000 pounds. The length of each truss between the centres of the end posts is 359.3 ft., making the dead load 1,901 pounds to the foot; a small allowance should be added to this for paint, wire railing, dust, ice, etc., increasing the dead load to 1,920 pounds per foot, which leaves a simple ratio to the assumed moving load of 2,240 pounds to the foot. The framing of the truss is such that the weight should be transferred to the turn-table only through the centre posts and cross-girders, the adjoining posts standing outside of the turn-table; the length of each arm should therefore be measured from the centre of the truss. The skew of the bridge makes the two arms of each truss of unequal length; the greater length, 182 feet, will be considered in these calculations, giving results which will be slightly excessive for the shorter arm. The total load per running foot on each truss is 2,080 pounds, of which $\tfrac{6}{13}$, or 960 pounds, is dead, and 1,120 is variable.

---

* The noted expansion of bottom chords from the coldest day in January, 1870, say 10° below zero, to the warmest day in June, 1870, say 104° above in the shade, was as follows:

133 ft. span	$\tfrac{7}{8}$ inches.	250 ft. span	$1\tfrac{1}{8}$ inches.
200 " "	$1\tfrac{1}{8}$ "	177 " "	$1\tfrac{1}{16}$ "

The draw was built open upon false-works extending up and down the river ; on the removal of the temporary supports it became a beam balanced upon its centre, and the same circumstance is repeated whenever the draw is open. The top chord is then in tension, and the bottom chord in compression, the strains tending to distort the beam in the manner shown in Fig. 1. The strains are greatest at the centre, where the moment of flexure is :—

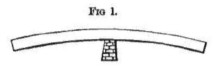

Fig 1.

$$M = -\frac{\overline{364}^2 \times 960}{8} = -15,899,520 \text{ pounds.}$$

and the moments throughout the beam will be proportional to the external ordinates of a parabola.* The chord strains will be equal to the moments divided by the depth of truss, the central strain being

$$-M \div 34.3 = 463,500 \text{ pounds.}$$

When the draw is closed, the wedge-plates are driven home under the end posts, giving the truss a bearing on the piers at each end. The power applied through these wedges is not sufficient to lift the ends of the draw, but merely to bring them in contact with the end supports, and prevent them from settling when a moving load passes. The whole weight continues, as before, to be carried by the pivot pier, and no change takes place in the strains.

When a train enters the draw, the variable load thrown upon one arm, will be borne in part by the wedges and in part by the turn-table. It will cause the loaded arm to deflect, and at the same time lift the end of the unloaded arm from its bearing on the wedges ; the distortions caused by these deflections will be similar to

Fig 2.

those shown in Fig 2, and will increase from the instant the engine enters the draw until the whole arm is covered by the load.†

The strains at the centre and in the unloaded arms remain unchanged ; those in the loaded arm become equal to the sum of the strains already existing, plus

---

* A diagram of the curves indicating the Moments in the Draw, is given on Plate XII ; the curve of Moments in the Open Draw is drawn in a plain black line.

† The passage of a heavy freight train has caused the further end of the draw to rise ¼ of an inch.

Engineered Irony

the strains due to the weight of the live load alone, the latter component strains being simply those caused by a load of the given intensity on an isolated span 182 feet long, and being tensile strains in the lower chord and compressive in the upper. The moment of flexure at the centre of the arm, due to the moving load only, is :—

$$\frac{182^2 \times 1120}{8} = 4,637,360 \text{ pounds.}$$

and the moments throughout the arm will be proportional to the internal ordinates of a parabola.* The resulting moments, the effect of the dead and live load combined, being at each point equal to the algebraic sum of the two moments already considered, are negative at the centre of the draw, decreasing in intensity, and finally becoming positive as they approach the end of the loaded arm.†

As the train advances upon the second arm, this arm will begin to deflect until it takes a bearing upon the wedges, while the strain over the pivot, which has hitherto remained unchanged, begins to increase. After an end bearing has been taken upon the wedges the second arm will continue to deflect at intermediate points until the entire draw is loaded, when the deflection of the two arms becomes symmetrical, and the distortions of the beam resemble those shown in Fig. 3. The strain

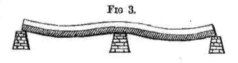

Fig 3.

over the pivot does not attain its greatest intensity until both spans are fully loaded ; it is then equal to the sum of the effects due to the dead load, which is carried wholly by the pivot pier, plus the effects of the live load, which is distributed according to the laws which govern a continuous beam resting on three supports. The moment due to the moving load, is at the centre the negative of that at the centre of one arm when loaded alone, being equal to

$$- 4,637,360 \text{ pounds.} ‡$$

this added to the moment due to the dead load gives for the maximum moment at the centre

$$- 20,536,880 \text{ pounds.}$$

---

* Shown on the diagram in a dotted line ....

† The curve of these moments is given on the diagram in a broken line — — — — —

‡ The curve of these moments is shown on the diagram in a broken line —...—...—...—...

Proceeding from the centre towards either end, these moments corresponding to a full load decrease in intensity till they become positive, being everywhere determined by the ordinates of a parabola.*

The central chord strains will be :—

$$20{,}536{,}880 \div 34.3 = 598{,}743 \text{ pounds.}$$

The greatest positive moments (compression above and tension below) occur when one arm only is loaded ; the greatest negative moments over the pivot pier occur when both arms are fully loaded, and those at the centre of each arm and near the ends of the draw take place when neither arm is loaded.†
The maximum moments at the end of each panel are given in the tables in Appendix G, the values being obtained by measuring on a diagram the ordinates of the curves of strain ; the corresponding chord strains are likewise given in the tables, their values being found by dividing the moments by the depth of truss at the corresponding point. To find the strains which actually take place in the chords, a correction must be made for the strain which is carried horizontally by the members of the web ; these corrected strains are also given in the tables, the tensile strains being distinguished by a negative sign ; as the strains carried horizontally by the web are all in tension, being carried by the diagonal ties, the actual amount of compression in the chord materially predominates over the tension.‡

The web is formed in two systems which are connected only through the chords. As each system is supposed to carry but half the load, in calculating the strains on their members the dead load will be assumed to be 480 pounds to the foot, and the live load 560 pounds. The strain in each post will be the same as the weight carried by the tie which depends from it ;§ the strain in each tie is equal to the weight carried, multiplied by the proper coefficient for the inclination of the tie.

---

* Shown on the diagram in a broken line —.—.—.—.—.—

† The resultant curve of maximum intensity of strain for all parts of the draw, without regard to the sign of the moments, is drawn on the diagram in a plain heavy line.

‡ Strictly speaking, plus the weight of the upper chord for one panel ; this weight however is small, and the calculation need not be complicated by it.

§ These corrected chord strains, expressed in tons of 2,000 pounds, are given on the skeleton diagram on Plate XII.

Engineered Irony

When the draw is swung or unloaded, each tie carries the load between the end of the draw and the foot of the tie next inside. The strains and weights which then occur are given in the tables in Appendix G.

When the draw is closed and loaded, the pivot pier carries the entire weight from the centre to that point on each arm where the moment of flexure has the greatest positive value; this point is half way between the point of reversal of the moments and the end of the draw; it is nearer the centre when only one arm is loaded than when both arms are loaded. The strains in the ties which carry this load to the central pier should therefore be calculated with reference to the case in which the whole draw is loaded, and the strains in those ties which carry this load towards the ends, with reference to the case in which but one arm is loaded.

When both arms of the draw are loaded the general equation for the moment of flexure is

$$M = -620\,l^2 + 1660\,l\,x - 1040x^2$$

l being the length of one arm and x the general abscissa. Differentiating :—

$$\frac{d\,M}{d\,x} = 1660\,l - 2080\,x$$

and the maximum value of M corresponds to

$$x = .79808\,l = 145.25.$$

The strains in the ties which carry this load towards the central post will therefore be calculated as if the centre of the truss was distant 145.25 feet from that post, making the equivalent length of span 290.5 feet.

The general equation of the strains in the web under the action of a moving load is

$$S = \frac{l^2\,w - x^2\,w'}{2\,l} - x\,w$$

in which l denotes the total length of span, w the dead load, and w' the moving load per foot. Substituting the values

$$(w = 480,\;w' = 560,\;\text{and}\; l = 290.5$$

this becomes

$$S = 69720 - .964\,x^2 - 480\,x$$

the value of x for each tie being the assumed length of beam (290.5 feet) minus the distance from the centre post to the foot of the tie next inside. The strains in the ties and posts, calculated by this formula, are given in the tables in Appendix G. As the practical centre is distant 145.25 feet from the centre post only when both spans are fully loaded, the results thus obtained are excessive for all but the centre ties, the others being most intensely strained under a partial load when the practical centre will be nearer the centre post and the practical length of beam less.

When but one arm is loaded the general equation for the moment of flexure is

$$M = -480\, l^2 + 1520\, l x - 1040\, x^2$$

the maximum value of M corresponding to

$$x = .73077\, l = 133.$$

The strains in the ties which carry their load to the end posts will therefore be calculated as if the centre of the beam was distant 133 feet from the centre post, or 49 feet from the end post, the equivalent length of span being 98 feet. Substituting $l=98$ in the general equation for the web strains, it becomes

$$S = 23520 - 2.857\, x^2 - 480\, x$$

the value of x for each tie being the assumed length of beam (98 feet) minus the distance from the end post to the foot of the next tie outside. The strain in the several ties and posts, calculated by this formula, are also given in Appendix G.*

The maximum strain on the centre post will be equal to the total dead and moving load between the points 145.25 feet on each side of that post, excepting the half panels adjoining, the weight of which is carried directly by the pivot. The strain is therefore

$$(290.5 - 15.5) \times 2080 = 572,000 \text{ pounds.}$$

In like manner the strain on each end post is found to be

$$(49 - 6.25) \times 2080 = 87,800 \text{ pounds.}$$

---

* The maximum web strains expressed in tons of 2,000 pounds are marked on the skeleton diagram on Plate XII. The practical centres used in calculating these strains are also designated on the same diagram.

Engineered Irony

The calculations of the contractors, by which the draw was framed, were made on the supposition that each arm acts as an isolated truss when the draw is closed ; the panel ties are therefore of the same size at the centre and ends of the draw. To make the actual strains agree with these calculations, the lifting jacks placed under the end posts must carry one-half the weight when the draw is fully loaded, the weight carried by each jack being

$$91 \times 2080 = 189{,}280 \text{ pounds.}$$

The weight carried by each end bearing, under the present arrangement of wedge plates, is

$$(182 - 145.25) \times 2080 = 76{,}440 \text{ pounds.}$$

making the weight which would actually be lifted by the jacks under each post, after closing the draw, 112,840 pounds.

CHAPTER VIII.

COST OF THE WORK.

THE cost of the bridge across the river, and of the approaches, as well as of the works incidental to its protection, and necessary to secure the permanency of the channel at this point, have been as follows to date (July 1, 1870):

BRIDGE ACROSS THE RIVER.

FOUNDATIONS.

Foundation South Abutment, on Rock	$458 51	
" Shore pillars, "	100 00	
" Pier No. 1, "	14,184 85	
" " No. 2 (Pivot), "	22,214 00	
" " No. 3, "	26,150 30	
" " No. 4, "	79,695 17	
" " No. 5, piles to rock,	37,873 05	
" " No. 6, on piles,	15,654 00	
" North Abutment, "	4,707 64	
		$201,037 52

MASONRY.

6,612 Cubic yards in Piers and Abutments	$116,023 10	
3,102 " " Concrete in foundations	31,470 00	
352 " " Rubble Masonry in foundation	3,168 90	
5,795 " " Riprap around Piers	24,429 75	
		$175,091 75

DRAW PROTECTION.

Crib, upper draw rest, below low-water line	$14,109 81	
Caisson, lower draw rest, " " " "	15,013 70	
Upper works of above and intermediate cribs and trusses	54,006 58	
Pontoon protection for boats at Pier No. 1	13,207 53	
		$96,337 62

Brought forward	$472,466 89

Carried forward... $472,466 89

OUTFIT.

Cost Steamer Gypsey..............................	$12,933 69		
Less proceed of sales..............................	4,000 00		
		$8,933 69	
Cost of barges and boats..........................	$13,949 67		
Less transferred to pontoons.......................	6,000 00		
		$7,949 67	
Tools, machinery, and equipment...................	$54,881 86		
Less proceed of sales..............................	9,073 53		
		$45,808 33	
Use of buildings, shanties, etc.....................		2,453 22	
			$65,144 91

MISCELLANEOUS EXPENSES.

Engineering.......................................	$23,269 00	
Soundings and borings in the river	3,224 71	
Superintendence and inspection....................	14,855 10	
Office expenses...................................	1,866 87	
Incidental expenses...............................	3,880 25	
Boating, ferriage, and cutting ice..................	53,039 35	
Contingencies (being cost of works of first caisson for Pier No. 4, up to the time of its wreck).........................	12,339 01	
		$112,474 29

SUPERSTRUCTURE.

Iron Span, No. 1, 71 feet....................	$5,335 98	
Wood and Iron, No. 2, 133 "	9,183 38	
Iron Draw, No. 3, 360 "	78,054 50	
Wood and Iron, No. 4, 200 "	18,639 14	
" " No. 5, 250 "	27,316 42	
" " No. 6, 200 "	18,639 14	
" " No. 7, 177 "	15,126 70	
Nicholson pavement, sidewalks, and hand-railing....	28,730 96	
Covering chords and painting bridge...............	7,918 47	
Toll-houses, telegraph lines, etc...................	2,503 91	
False-works for raising superstructure,.............	21,142 87	
Street rail track on bridge........................	3,494 22	
		$236,085 69
Total cost of bridge proper.....................		$886,171 78

APPROACHES AND RIVER PROTECTION.

Southern Approach (Kansas City side)...............	$63,340 09	
Northern Approach (including trestle)...............	51,723 22	
Depot grounds....................................	1,003 68	
Protection of river bank...........................	90,938 81	
		$207,005 80
Total expenditure..............................		$1,093,177 58

It will be noted from the above, that the cost of the bridge proper was $886,171.78, or $637.08 per lineal foot. While this does not materially differ in the aggregate from the cost of similar structures over the Mississippi River, yet the cost per running foot is nearly twice as great, in consequence of the deep and difficult nature of the foundations, and the greatly increased height of the piers.

If, however, the cost be referred, as is sometimes done, to the area enclosed between the top of the bridge superstructure, and the bottom of its foundations, it will be found, that as this cross section measures 147,020 square feet, the cost has been $6.03 per square foot of the area so enclosed, which will compare favorably with similar works.

It must be remembered that all the foundations have been put in for a double-track bridge, up to low water, so that in the not improbable event that the traffic shall require it, the capacity of the bridge can be doubled at no very great cost.

No separate account was kept of the additional cost occasioned by the accommodation of the roadway to the wagon traffic, but it may be stated in round numbers at about $40,000.

It is not expected that any further expenditures will be required for the next ten or fifteen years, save for the maintenance of the river protection, upon which depends the permanence of the river bank and channel, as well as the harbor at Kansas City, and for the renewal of the upper works of the draw protection, which are of wood.

The wood and iron combination of the fixed spans being experimental, so far as duration is concerned, there are no data at hand upon which to predicate an estimate of their life. It may be stated, however, that a superstructure all of iron would have cost $72,000 more, and that as all the wooden parts can be replaced for $23,000, it follows that the compound interest alone at 7 per cent. upon the sum saved, will renew all the perishable portions of the superstructure with wood every four years, while at the end of fifteen years it would be sufficient to renew them all with iron.

It is believed that the combination adopted, and the pains which have been taken for its preservation, may insure for it even a longer life than this.

Engineered Irony

Wooden bridges well housed in, have been known to last sixty and even a hundred years, and it is hoped that the wooden parts of this may endure, with occasional repairs, even twenty or thirty years. This point, however, must be decided by the sole test of experience.

The following is a synopsis of the materials in the superstructure :

Timber in floor of iron span	71 feet	7,684	F. B. M.
" " truss, etc.	133 "	35,739	"
" " pivot span	360 "	26,025	"
" " truss, etc.	250 "	101,688	"
" " " " 2 spans	200 "	156,430	"
" " " "	177 "	57,854	"
" " roofing and flooring		210,876	"
" " vertical bracing		8,202	"
Total timber		604,498	"

Wrought iron in iron span	71 feet	32,165	lbs.
" " " truss	133 "	44,053	"
" " pivot span	360 "	495,575	"
" " truss	250 "	147,432	"
" " " 2 spans	200 "	178,898	"
" " "	177 "	72,969	"
" " anchor rods, pins, etc.		7,200	"
Total wrought iron		978,292	

Castings in iron span	71 feet	4,328	lbs.
" " truss	133 "	27,138	"
" " pivot span	360 "	122,041	"
" " truss	250 "	70,646	"
" " " 2 spans	200 "	108,238	"
" " "	177 "	49,491	"
" " anchors, braces, etc.		1,700	"
Total cast-iron		383,582	

The position of Kansas City, being as it were on the frontier, made skilled labor both expensive and difficult to obtain during the first year. After this time, the increasing population of the town, and the attention attracted by these works, furnished all that was required.

The following were the average wages paid :

Superintendents of departments..............from $200 to $250 per month.
Foremen blacksmiths, carpenters, and laborers. "   3 75 "   5 00 per day.
Blacksmiths................................ "   3 50 "   3 75   "
Helpers.................................... "   2 50 "   2 75   "
Carpenters .... "   3 00 "   3 25   "
Stone-cutters and masons................... "   4 50 "   5 50   "
Laborers .... "   2 00 "   2 25   "
Pile-drivers and boatmen................... "   2 25 "   2 50   "

The following were the leading prices of materials :

Oak timber and plank..........................$35 to $40 per M. B. M.
Pine lumber....................................... 45 "   60 "   "
Cotton-wood lumber .... 20 "   25 "   "
Piles, oak, elm, and sycamore.................... 15 cts. to 25 cts. per foot.
Bar iron.. 4¼ "   " 4¾ " per lb.
Cast iron, to order............................... 6 "   " 8 "   "
Bolts, and miscellaneous iron work................ 6 "   " 15 "   "
Hydraulic cement.................................. $3 25 " $3 75 per bbl.
Riprap stone .... 2 00 " 2 50 per c. yd.

Upon looking back over the methods adopted for founding each pier, and the general plans which were carried out, the engineers see no reason to alter their judgment of the appropriateness of each to the particular location selected. Yet they are conscious of many possible changes and improvements in the details which would have materially hastened and cheapened the work.

Thus, at Pier No. 1, it would still be wished to sink a bottomless caisson, protected by a water deadener, and lay bare the rock; but a yielding cushion fastened to the lower edge, would probably enable the making of a water-tight joint with much less time and expense.

At Pier No. 2, it is still thought judicious to found in one mass, and to make the current perform the excavation; but the device alluded to above, and the hanging of the false bottom lower down, would probably have avoided the cone of sand which remained on the rock, causing great delay and expense; and it might have been possible to lay bare the rock.

At Piers Nos. 3 and 5, greater weight given to the caissons would undoubtedly have hastened their descent, and lessened the amount of the excavation.

At Pier No. 4, the most expensive of all, the plans would not be materially changed ; but details would be altered, so that the pier could be built, and the foundation completed much more rapidly than it was.

At Piers Nos. 6 and 7, the plans would not be materially changed.

All the delays, difficulties and failures which took place were directly owing to the violence of the current, and its capacity for rapid scour. The precautions and watchfulness which these required, both by night and by day, were endless, and not always successful.

The moods of the river were constantly changing, and its bottom and banks of most unstable regimen, thus causing no little anxiety and expense, while the absence of precedents in this kind of work, in this country, left the engineers to depend mostly upon their own resources.

It is hoped that this imperfect relation of the experience acquired upon this novel work may be of profit to others engaged in similar undertakings.

APPENDIX.

Engineered Irony

APPENDIX A.

EXTRACTS FROM AN ACT TO INCORPORATE THE KANSAS CITY, GALVESTON, AND LAKE SUPERIOR RAILROAD COMPANY.

Be it enacted by the General Assembly of the State of Missouri, as follows :

SEC. 1. That a Company is hereby incorporated, to be called the Kansas City, Galveston, and Lake Superior Railroad Company, the stock whereof shall be six millions of dollars, to be divided into shares of one hundred dollars each, the holders whereof, their successors or assigns, shall constitute and be a body corporate and politic, in law and in fact, by the name and style aforesaid, and by that name shall have perpetual succession; may sue and be sued, plead and be impleaded, defend and be defended against; may make and use a common seal, and break and change the same, and shall be able, in law and equity, to make contracts; may make, hold, use, possess, and enjoy the fee-simple, or other titles, in and to any real estate, and may sell and dispose of the same ; may make by-laws, rules and regulations proper for carrying into effect the provisions of this Act, not repugnant to the Constitution or laws of the United States, or of this State, and shall have the usual and necessary powers of corporation for such purposes.

*　　*　　*　　*　　*　　*　　*　　*　　*　　*

§ 7. Said Company shall have full power to survey, mark, locate, construct, maintain, and operate a railroad from the City of Kansas, in Jackson County, by the most direct and practicable route in the direction of Galveston, in the State of Texas, or to intersect any road or roads now being constructed, or to be constructed, by the States of Texas or Arkansas, or by any company or companies, which are or may be chartered by either or both of said States ; also, to construct said railroad north from the City of Kansas, or from the north bank of the Missouri River, opposite said City of Kansas, by the most direct and practicable route to the north boundary of the State, in the direction of Fon du Lac on Lake Superior ; and for that purpose may hold a strip of land not exceeding one hundred feet in width, with as many tracks as the said President and Directors may deem necessary ; *Provided*, that in passing hills or valleys, the said Company are authorized to extend said width, in order to effect said object, and may, also, hold sufficient land for the erection of depots, warehouses, and water stations, and may select such route as may be deemed most advantageous, and may extend branch railroads to any point in any of the counties through which the said railroad may be located.

*　　*　　*　　*　　*　　*　　*　　*　　*　　*

§ 11. Said Company may build said road along or across any State or county road, or street or wall of any town or city, or over any stream or highway ; but whenever said railroad shall cross any State or county road, said Company shall keep good and sufficient causeways or other adequate facilities for crossing the same; and said railroad shall not be so constructed as to prevent the public from using any street, road, or highway, along or across which it may pass; and when said railroad shall be built across any navigable stream, said Company shall erect a bridge sufficiently high on which to cross, or shall construct a drawbridge, so that in no case shall the free navigation of such stream be obstructed. When any person shall own land on both sides of said road, said Company, when required so to do, shall make, and keep in good repair, one causeway or other adequate means of crossing the same.

*   *   *   *   *   *   *   *   *   *

§ 16. Said Company shall have power to extend, construct, maintain, and operate their said railroad or branches, beyond the limits of this State, and are hereby authorized to make contracts for the same, and shall have power to contract for, and construct, all necessary bridges over navigable streams, so the same may in nowise interfere with the free navigation of the same.

*   *   *   *   *   *   *   *   *   *

This Act to take effect, and be in force, from and after its passage.

Approved February 9, 1857.

APPENDIX B.

AN ACT TO INCORPORATE THE KANSAS CITY BRIDGE COMPANY.

Be it enacted by the General Assembly of the State of Missouri, as follows:

Sec. 1. That R. T. Van Horn, M. J. Paine, A. J. Lloyd, and David E. James, their associates, assigns, and successors, are hereby constituted a body corporate and politic, by the name of the Kansas City Bridge Company, and shall have the exclusive right and privilege of constructing a bridge at or near the City of Kansas, over and across the Missouri River, for the term of twenty years, and no other person or company whatsoever shall construct any other bridge for or within the distance of one mile from said bridge during the term of said twenty years, and said Company, by their corporate name, shall be capable in law of purchasing, taking, holding, using, selling, pledging, conveying, and disposing of real estate or other property, whether personal or mixed, so far as the same may be necessary for the purposes herein mentioned; may have a common seal, may sue and be sued, plead and be impleaded, defend and be defended against, contract and be contracted with; may make such by-laws, rules and regulations, appoint such officers, agents, and servants, and generally do all such acts and things not inconsistent with the laws and Constitution of the State of Missouri, and of the United States, as may be requisite and proper for the due execution and management of the work herein proposed to be done and for conducting the business of said Company.

§ 2. All the powers of said corporation shall be exercised by a board of directors and such officers and agents as they may elect and appoint. Said board shall consist of nine directors, who shall be elected annually by a majority in interest of the stockholders, present and voting, in person or by proxy, at such time and place as shall be prescribed by the by-laws, and who shall hold their offices until their successors are elected and qualified. Previous to such election the corporators herein named shall constitute the board of directors, three of whom shall constitute a quorum to do business. They may at any time after the passage of this Act cause books to be opened for subscriptions to the capital stock of the said Company at such times and places, and in such manner, as may be by them prescribed.

§ 3. The capital stock of said Company shall be one million of dollars, which shall be divided into ten thousand shares of one hundred dollars each, and when two hundred and fifty thousand dollars shall be subscribed the corporators shall call a meeting of the stockholders for the purpose of electing a board of directors, at such time and place as

Engineered Irony

they may fix and appoint, of which election thirty days notice shall be given in a news-paper published in the City of Kansas.

§ 4. The said Company shall have the same power to condemn and acquire title to lands necessary for the construction of said bridge and for the approaches to it from public highways, and the same power to take materials from lands in the neighborhood for the construction thereof, compensating the owners thereof, and the proceedings therein shall be conducted in the same manner as is provided in the Act to authorize the formation of railroad associations and to regulate the same, approved December 13, 1855, and the said Company shall also have the right to protect the banks of the river above and below the bridge, so far as may be necessary to keep the channel within the openings of the bridge for the passage of vessels, and for that purpose to acquire, con-demn, and take lands and materials in the manner aforesaid.

§ 5. The said bridge shall be constructed so as not to prevent the passage of steam-boats or other vessels in the navigation of said river, and said bridge may be built so as to admit a railroad track for the passage of cars and trains, as well as for a common wagon way, and for a foot passenger way. It shall be properly attended and managed, so as to afford safe and easy passage for all persons and property, and every railroad train drawn by steam power shall, on approaching said bridge, stop its speed and come to a stand-still on the bank before entering upon the passage of the bridge.

§ 6. When said bridge is completed the said Company shall be entitled to demand and receive tolls for crossing the same, and to fix the rates of toll, of which a schedule shall be kept conspicuously posted at each end of the bridge, which rates shall be as follows, and shall never exceed the same, to wit : For each foot passenger, five cents; for every person on horseback, twenty cents ; for every gig, buggy, or other travelling carriage, drawn by one animal, forty cents ; for every cart or wagon drawn by one animal, forty cents ; for every wagon or cart drawn by two animals, sixty cents ; for every cart or wagon drawn by three animals, eighty cents ; for every cart or wagon drawn by four animals, one hundred cents ; for every cart or wagon drawn by more than four animals, ten cents extra for each animal ; for every pleasure carriage drawn by two animals, sixty cents ; for every pleasure carriage drawn by four animals or more, one hundred cents ; for each head of cattle, horses, mules, or other working animal, ten cents ; for each head of sheep or swine, three cents ; and said bridge company may permit any railroad com-pany to extend their railroad track over said bridge upon such terms as may be agreed upon by said bridge company and such railroad companies.

§ 7. The corporation hereby created shall be exempt from the operation of sections six, thirteen, fourteen, fifteen, sixteen, eighteen and twenty, of article one of " An Act con-cerning Corporations," approved November 23, 1855.

§ 8. For the purposes of taxation, all property owned by this corporation shall be taken and deemed as personal property, in the shape of stock in the hands of stockholders, and the same shall be assessed and taxed as stock only.

§ 9. All acts and parts of acts whatsoever, inconsistent or in conflict with the provisions of this act, are hereby repealed.

§ 10. This act shall be taken as a public act, and shall be in force from and after its passage.

Approved February 20, 1865.

Engineered Irony

APPENDIX C.

AN ACT TO AMEND AN ACT ENTITLED "AN ACT TO INCORPORATE THE KANSAS CITY, GALVESTON, AND LAKE SUPERIOR RAILROAD COMPANY," APPROVED FEBRUARY 9, 1857.

Be it enacted by the General Assembly of the State of Missouri, as follows:

SEC. 1. That any deed of trust thereafter to be made and executed by said Railroad Company, now known as the Kansas City and Cameron Railroad Company, to secure the payment of bonds sold or to be sold by it to procure money for the completion of its road and appurtenances, shall be a valid lien upon all property described in said deed of trust, and upon the entire line of said road and its appurtenances and franchises, although the said property might not have been obtained or said road completed at the time of the execution of said mortgage or deed of trust; and the said bonds to be issued thereunder may bear such rate of interest, and be sold for such price, as may be deemed expedient by the Board of Directors of said Company, and it shall be competent and lawful for the said Company to make and enter into such contract with the trustees in such deed of trust as will secure the just and true application of all moneys raised under it, and of all other funds of the Company, to the work of construction of said railroad, and to secure the same under said deed of trust as a security for the payment of said bonds.

SEC. 2. It shall be lawful and competent for said Company to make such arrangement with any other railroad company to furnish equipments, and to run and manage its railroad, as it may deem expedient and find necessary, or to lease the same, or to consolidate it with any other company upon such terms as may be deemed just and proper.

SEC. 3. For the purpose of adding to the safety of its bonded debt, it shall be competent and lawful for the holders of its bonds to vote at all elections of the Company, and to be represented at all meetings of the stockholders, and vote either in person or by proxy. The holder of each one hundred dollars of such bonds shall be entitled to the same vote and representation as the holders of each one hundred dollars of stock of said Company; and it shall be competent and lawful for said Company to provide for a registration of such bonds in such manner that they may be made payable to order, and the circulation restricted at the pleasure of each holder; and such rules and regulations shall be made by the said Board of Directors as will accomplish this object, and to secure the safety of said bonds as may thus be possible against theft or other losses thereof.

Sec. 4. The said Railroad Company shall have the same authority, rights and powers as are conferred upon the Kansas City Bridge Company, incorporated by an Act of the General Assembly of Feb. 20, 1865, and may, in connection with its railroad bridge, erect a bridge for the passage of teams, carriages, and foot passengers, and shall have the same right and authority to receive compensation therefor as are granted to the said Kansas City Bridge Company; and all railroad companies whose roads shall terminate at or near such bridge on either side of the Missouri River, or which shall construct a branch road to such bridge, shall have the right to run their cars and engines on and over such bridge, at such times and on such terms as may be agreed on between the companies, respectively; and if such companies shall not agree on such terms, then on such terms as shall be prescribed by the Governor of this State.

Sec. 5. The North Missouri Railroad Company shall have the privilege of laying their track over the right of way of the Kansas City and Cameron Railroad Company, where it passes the bluff at Randolph, on the Missouri River, and thence to a point on said river opposite to the City of Kansas, upon condition that the said Company shall not lay their track within nine feet of or in any manner so as to interfere with the Kansas City and Cameron railroad track as at present located; nor shall it cross the said track until within half a mile of the railroad bridge now in course of construction at the City of Kansas, unless the two companies otherwise agree; and in case the North Missouri Railroad Company do not construct and terminate their road at the City of Kansas, they shall pay the Kansas City and Cameron Railroad Company a just valuation of the right of way aforesaid, which value, if not mutually agreed to, shall be determined by three railroad experts, each company to select one, and the two so chosen to select the third; or in case the two companies shall agree to the joint use of the track of the Kansas City and Cameron Railroad, from their eastern intersection to the City of Kansas, but shall disagree as to the annual compensation to be paid the said Kansas City and Cameron Railroad Company for the use of their track, then the amount to be so paid by the North Missouri Railroad Company shall be determined by three railroad experts, to be chosen as provided, in case of disagreement as to right of way.

This Act to take effect, and be in force, from and after its passage.

Approved March 11, 1867.

17

APPENDIX D.

Statement of Traffic on the Kansas City Bridge, from July 13th, 1869, to February 28th, 1870, in both directions.

MONTHS.	TOLLS RECEIVED FROM HIGHWAY TRAFFIC.	LOCOMOTIVES.				BOX CARS.				FLAT CARS.				COACHES.			
		WITH TRAINS.		WILD.		LOADED.		EMPTY.		LOADED.		EMPTY.		PASSENGER.		BAGGAGE.	
		H. & St. Jo. R.R.	N. Mo. R.R.	H. & St. Jo. R.R.	N. Mo. R.R.	H. & St. Jo. R.R.	N. Mo. R.R.	H. & St. Jo. R.R.	N. Mo. R.R.	H. & St. Jo. R.R.	N. Mo. R.R.	H. & St. Jo. R.R.	N. Mo. R.R.	H. & St. Jo. R.R.	N. Mo. R.R.	H. & St. Jo. R.R.	N. Mo. R.R.
July 13th to 31st........	$113 75	164	118	36	29	435	238	268	213	83	69	53	43	164	273	64	101
August 1st to 31st........	320 65	273	226	147	47	835	465	417	323	154	116	173	133	388	285	115	163
September 1st to 30th.......	403 95	300	240	263	162	1130	476	400	305	191	152	159	132	429	345	125	168
October 1st to 31st........	917 85	315	262	169	92	1405	555	356	356	81	187	184	227	607	388	167	168
November 1st to 30th.......	941 47	325	265	66	17	1069	512	982	333	312	268	283	234	517	399	163	168
December 1st to 31st.......	1190 40	304	269	27	20	1183	525	844	417	439	454	440	459	509	382	190	245
January 1st to 31st........	889 45	283	260	12	12	891	575	643	318	255	486	132	465	478	312	167	173
February 1st to 28th.......	868 50	302	225	21	12	1153	596	920	495	127	244	126	287	514	276	157	153
Total........	$5706 02	2266	1865	741	391	8101	3942	4730	2760	1642	1976	1550	1980	3806	2660	1148	1339

Engineered Irony

APPENDIX E.

Tables showing Progress of Sinking Pier No. 4, with Sounding,

Date.	Shift.	Hours Work of Dredges					Cubic Yards Excavated.	Reading of Gauges									Depth Sunk.	Elevation Cut'ng Edge.
		1	2	3	4	Total.		1	2	3	4	5	6	7	8	Aver.		
Dec. 28.	Day.	4½	4	4	4¼	17	20	....	....	....	....	....	....	....	....	....	18"	....
" 29.	"	7½	7	7½	7½	29½	50	20.6	20.8	21.	21	21.	21.1	21.	21.3	20 98	6½	88 02
" 30.	"	8	6	0	4	18	35	21.	21.3	21.45	21.4	21.3	21.35	21.25	21.3	21 29	7	87.71
" 31.	"	7	7½	3	0	17½	20	21.6	21.9	21.7	21.6	21.3	21.35	21.25	21.75	21 57	3½	87.43
Jan. 1.	"	6	0	7½	5	18½	58	22 8	23.	23 35	23.5	23 5	23 5	23.6	23.75	23 37	20	85 63
" 2.	"	4	5	0	0	9	18	23.7	23 7	23.7	23.6	23.5	23.6	23.9	23.9	23.7	4	85.3
" 4.	"	10	10	7	9	36	120	26.	26 2	26	25 7	25.35	25 7	25.8	26.1	25.86	26	83.14
" 4.	Night.	7½	5	6	6¾	25¼	90	27 25	27.6	27.55	27.5	27 2	27 4	27.44	27.45	27.42	18½	81.56
" 5.	Day.	4	3	11	2	20	56	28.55	28.75	28.5	28.2	27 9	28 2	28.4	28 65	28 41	11¾	80.50
" 5.	Night	9	9½	8	7¼	33¾	100	29.9	30.2	30 25	30.25	30.05	30 2	30.25	30.15	30.16	21	78.84
" 6.	Day.	0	0	0	0	0	0	....	....	....	....	....	....	....	....	....	0	....
" 6.	Night	4½	3¾	4¾	4¾	17¾	65	31 1	31.3	31 43	31.47	31.23	31.35	31.35	31.25	31.31	13¾	77.69
" 7.	Day.	0	0	0	0	0	0	....	....	....	....	....	....	....	....	....	9	....
" 8.	"	0	0	0	0	0	0	....	....	....	....	....	....	....	....	....	0	....
" 9.	"	0	0	0	0	0	0	....	....	....	....	....	....	....	....	....	0	....
" 11.	"	0	0	0	0	0	0	31.4	31.75	....	31.2	31 6	31.63	31.6	31.55	31.62	3¾	77.38
" 12.	"	0	0	0	0	0	0	....	....	....	....	....	....	....	....	....	0	....
" 13.	"	0	0	0	0	0	0	31.65	31.8	31.9	32.	31.8	31.75	31.75	31 65	31.79	1¾	77 21
" 13.	Night	5¾	3½	4½	2	15¾	67	32.95	33.05	33 05	33 05	32.85	32.75	32.75	32 8	32.91	13½	76 09
" 14.	"	5½	5½	5½	5½	22	55	33 85	33.9	33 95	33.95	33.85	33.8	33.85	33.8	33 87	11¾	75.13
" 15.	"	0	0	0	0	0	0	33.85	33.9	34.	34.	33 9	33.85	33 85	33.85	33.9	0	75.1
" 16.	"	0	0	0	0	0	0	....	....	....	....	....	....	....	....	....	0	....
" 17.	"	0	0	0	0	0	0	....	....	....	....	....	....	....	....	....	0	....
" 18.	"	0	0	0	0	0	0	....	....	....	....	....	....	....	....	....	0	....
" 19.	"	2¾	4	4¾	4¾	16¼	85	35.4	35.43	35 37	35.25	35.2	35 15	35 3	35.4	35 31	17	73.60
" 20.	"	0	0	0	0	0	0	36.37	36 4	36 35	36 25	36.2	36 15	36 3	36.35	36.3	11¾	72.7
" 21.	"	0	0	0	0	0	0	36.4	36.45	36.5	36 45	36.4	36.35	36 4	36.4	36.42	1½	72.56
" 22.	"	0	0	0	12	12	40	....	....	....	....	....	....	....	....	....	0	....
" 23.	Day.	0	0	0	0	0	0	36.8	36 8	36.9	36 85	36.85	36.85	36.85	36.8	36 83	5½	72.17
" 23.	Night	4	8	5½	4	21½	70	38.1	38.05	37 9	37.75	37 85	37 6	37.85	38 05	37.75	12½	71 25
" 25.	"	0	7	7	5½	19½	56	38 5	38.45	38.5	38 45	38 5	38 5	38.5	38.45	38.47	9	70.31
" 26.	"	2	3	3	1	9	50	39 3	39 3	39 27	39 17	39.25	39 1	39 25	39.3	39.23	9	69.77
" 27.	"	6	6¼	7	5¾	25	150	41.7	41.75	41.8	41.7	41.62	41 73	41.82	41.8	41 75	30¾	67.25
" 28.	"	0	0	0	0	0	0	....	....	....	....	....	....	....	....	....	0	....
" 29.	"	3½	4	4	3	14½	90	43 33	43.	43.1	42.95	43.1	43.25	43.1	43.1	43 11	16¾	65 89
" 30.	"	2½	3	3¼	3	11¾	85	43.9	43 97	44.05	44.05	44.1	44.15	44.1	44.1	44 05	11½	64 95
" 31.	"	3¾	3½	1¼	2½	10¾	75	45.15	45 33	45.27	45.23	45.3	45.4	45 35	45 35	45.28	14¾	63.72

Weights, etc., compiled from Journal kept on the Works.

SOUNDINGS.									Elevation Sand.	Height Water.	DISPLACEMENT.			WEIGHT.		SAND IN CONTACT.		Weight per Sq. Ft.	Friction per Sq. Ft.	DATE.
1	2	3	4	5	6	7	8	Av.			Ver.	Cu. Ft.	Pounds.	Total.	Effective.	Ver.	Sq. Ft.			
....	....	....	....	....	....	....	....	....	101.3	....	....	....	....	1,164,400	....	....	....	....	....	Dec. 28.
8.5	9.5	9.5	10.5	9.5	11.	10.5	10.5	9.6	91.8	101.4	13.4	8,688	541,819	1,164,400	622,581	3.8	601.9	1036.3	30.4	" 29.
8.5	9.5	9.5	10.5	9.5	11.	11.5	10.5	10.1	91.6	101.7	14.	9,365	584,376	1,164,400	580,024	3.9	617.8	....	....	" 30.
10.5	9.5	10.	....	10.	....	11.5	....	10.3	91.2	101.5	14.1	9,478	591,427	1,164,400	572,973	3.6	601.9	....	....	" 31.
9.5	....	10.5	....	8.5	....	11.5	....	10.	91.7	101.7	16.1	11,732	732,077	1,225,200	493,123	6.1	966.2	....	....	Jan. 1.
9.	9	10.	10.	8	10.	12.	11.	9.9	92.1	102.	16.7	12,405	774,072	1,278,000	503,928	6.8	1077.1	467.8	54.4	" 2.
8.	.8.5	8.5	9.	7.5	8.5	10.5	11.5	9.	92.4	101.4	18.3	14,187	885,269	1,676,400	791,131	9.3	1473.1	....	....	" 4.
									92.5	101.6	20.	16,063	1,002,331	....	....	10.9	1726.6			" 4.
7.5	7.5	7.5	9.	9.	9.5	11.5	10.5	9.	92.6	101.6	21.	17,162	1,070,909	1,874,400	805,491	12.	1900.8	423.8	96.	" 5.
8.	9.	10.	8.5	6.	9.	12.	10.5	9.1	92.6	101.7	22.9	19,232	1,200,077	....	....	13.0	2185.9	....	....	" 5.
....	....	....	....	....	....	....	....	....	....	101.8	....	....	....	2,026,800	....	....	....	....	....	" 6.
....	....	....	....	....	....	....	....	92.5	101.9	24.2	20,636	1,287,686	....	....	14.8	2344.3	....	....		" 6.
....	....	....	....	....	....	....	....		102.2	....	....	....	2,255,600	....	....	....	....	....		" 7.
....	....	....	....	....	....	....	....		102.7	....	....	....	2,406,000	....	....	....	....	....		" 8.
....	....	....	....	....	....	....	....		103.6	....	....	....	2,466,800	....	....	....	....	....		" 9.
....	....	....	....	....	....	....	....	91.	102.	24.6	21,067	1,314,581	....	....	13.6	2154.2	....	....		" 11.
....	....	....	....	....	....	....	....		101.7	....	....	....	....	....	....	....	....	....		" 12.
....	....	....	....	....	....	....	....	91.	101.3	24.1	20,529	1,281,010	2,556,800	1,275,790	13.8	2185.9	582.4	110.4		" 13.
....	....	....	....	....	....	....	....	91.	101.5	25.5	22,029	1,374,610	....	....	14.9	....	....	....		" 13.
14.8	14.	14.	12.	11.5	6.2	7.3	6.	11.	91.	102.	26.9	23,722	1,480,253	....	....	15.9	2518.6	....	....	" 14.
14.4	14.4	15.	12.5	12.3	7.	7.8	8.5	11.5	91.3	102.8	27.7	24,698	1,541,155	....	....	16.2	2566.1	....	....	" 15.
15.3	12.	12.	13.	12.5	6.5	8.5	9	11.1	92.5	103.6	....	....	....	....	....	....	....	....	....	" 16.
15	14.	12.	12.	12.	6.	7.5	9.	10.9	93.3	104.2	....	....	....	....	....	....	....	....	....	" 17.
14.	13.	9.5	10.	9.	4.	6.	7.5	9.2	93.3	103.5	....	....	....	....	....	....	....	....	....	" 18.
11.5	10.5	7	6.5	9.	3.	4.	6.5	7.3	95.5	102.8	29.1	26,406	1,647,734	3,006,800	1,359,066	21.8	3453.1	393.6	174.4	" 19.
12.5	11.	11.	7.	9.	3.5	5.	8.	8.4	94.1	102.5	29.8	27,260	1,701,024	....	....	21.4	338.98	....	....	" 20.
10.5	9	4.5	6.5	8.	3.	4.5	6	6.5	95.5	102.	29.4	26,772	1,670,578	....	....	22.9	3627.4	....	....	" 21.
10.5	9.5	5.	6.5	7.5	6.5	4.5	8.5	7.3	94.6	101.9	....	....	....	....	....	....	....	....	....	" 22.
10.	9	4.5	7.5	7.	7.	3.5	6.	6.8	94.9	101.7	29.5	26,894	1,678,186	....	....	22.7	3595.7	....	....	" 23.
....	....	....	....	....	....	....	....		94.9	101.7	30.5	28,114	1,754,314	....	....	23.5	3722.4	....	....	" 23.
10.	9.	4.	7.	6.	8.	5.	6.	6.9	94.8	101.7	31.2	28,968	1,807,603	....	....	24.3	3849.1	....	....	" 25.
....	....	....	....	....	....	....	....	6.9	94.5	101.4	31.6	29,456	1,838,054	....	....	24.7	3912.5	....	....	" 26.
9.5	9.	5.	9.	8.5	9.	4	5.	7.4	93.9	101.3	34.1	32,506	2,028,374	3,682,000	1,653,626	26.7	4229.3	391.	213.6	" 27.
....	....	....	....	....	....	....	....		101.5	....	....	....	....	....	....	....	....	....		" 28.
10.	9.5	6.	9	9	9.	6.	6.	8.2	93.6	101.8	35.9	33,722	2,104,253	....	....	27.7	4387.7	....	....	" 29.
10	9.	6	9.	8.5	9.	6.	6.	7.9	93.8	101.7	36.8	34,700	2,165,280	....	....	28.9	4577.8	....	....	" 30.
10.	9.	6.	9.5	8	9.	6.	6.	7.9	93.7	101.6	37.9	37,142	2,317,659	....	....	30.	4752.	....	....	" 31.

APPENDIX E—

Date.		Shift.	Hours Work of Dredges.					Cubic Yards Excavated.	Reading of Gauges.									Depth Sunk.	Elevation Cut'ng Edge.
			1	2	3	4	Total.		1	2	3	4	5	6	7	8	Aver.		
Feb.	4.	Night	5¾	6¾	6¾	4½	23¾	150	47.45	47.4	47.4	47.4	47.47	47.55	47.5	47.5	47.45	26″	61.55
"	5.	"	9½	7½	7	2½	26½	50	47.95	47.9	47.9	47.9	48.	48.05	48	48.	47.96	6	61.04
"	6.	Day.	0	0	0	0	0	0	48.07	48.	48.	48.	48.07	48.15	48.15	48.15	48.07	1¼	60.93
"	6.	Night	10	9½	7½	10	37	10	48.17	48.1	48.15	48.2	48.1	48.3	48.25	48.25	48.19	1¾	60.81
"	8.	"	10	10	10	10	40	12	48.45	48.4	48.4	48.4	48.57	48.65	48.52	48.5	48.46	3	60.54
"	9.	....	8	8	8	8	32	30	49.33	48.96	48.96	48.96	49.03	49.1	49.1	49.1	49.07	6¾	59.93
"	12.	....	7	6	5	5	23	18	....	49.88	....	49.75	....	49.9	....	50.25	49.9	7	59.1
"	15.	....	....	....	....	....	....	10	....	....	....	....	....	....	....	....	....	0	....
"	17.	....	....	....	....	....	....	....	....	....	....	....	....	....	....	....	....	0	....
"	19.	....	....	....	....	....	....	....	....	....	....	....	....	....	....	....	....	0	....
"	20.	....	....	....	....	....	....	....	....	50.4	....	50.	....	50.12	....	50.4	50.23	3¾	58.77
"	22.	....	....	....	....	....	....	....	....	....	....	....	....	....	....	....	....	0	....
"	23.	....	....	....	....	....	....	....	....	....	....	....	....	....	....	....	....	0	....
"	24.	....	....	....	....	....	....	....	....	50.75	....	50.45	....	50.55	....	50.7	50.56	4¾	58.44
"	25.	....	....	....	....	....	....	....	....	51.	....	50.77	....	50.9	....	51.	50.92	3¾	58.08
"	26.	Day	....	....	....	....	....	....	....	51.27	....	51.03	....	51.15	....	51.3	51.19	3¼	57.81
"	26.	Night	....	....	....	....	....	....	....	51.42	....	51.15	....	51.26	....	51.47	51.32	1¾	57.68
"	27.	Day.	....	....	....	....	....	....	....	51.45	....	51.15	....	51.17	....	51.5	51.32	0	57.68
Mar.	1.	Day & Night	....	....	....	....	....	....	....	51.62	....	51.3	....	51.2	....	51.7	51.48	2	57.52
"	2.	"	....	....	....	....	....	....	....	51.75	....	51.4	....	51.4	....	51.8	51.59	1¼	57.41
"	3.	"	....	....	....	....	....	....	....	51.85	....	51.52	....	51.47	....	51.85	51.67	1	57.33
"	4.	"	....	....	....	....	....	....	....	51.92	....	51.57	....	51.47	....	51.92	51.72	⅝	57.28
"	5.	"	....	....	....	....	....	....	....	52.1	....	51.65	....	51.7	....	52.1	51.89	2	57.11
"	6.	"	....	....	....	....	....	....	....	52.17	....	51.8	....	51.75	....	52.17	51.97	1	57.03
"	8.	"	....	....	....	....	....	....	....	52.26	....	52.	....	51.97	....	52.3	52.12	2	56.88
"	9.	"	....	....	....	....	....	....	....	52.4	....	52.15	....	51.15	....	52.49	52.28	1¾	56.72
"	10.	"	....	....	....	....	....	....	....	52.6	....	52.22	....	51.25	....	52.6	52.42	1⅝	56.58
Divers work..		....	....	....	....	....	171	200	....	....	....	....	....	....	....	....	....	....	....
Total....		....	....		....	....	773	1985	....	....	....	....	....	....	....	....	....	34′ 5½″	....

Continued.

1	2	3	4	5	6	7	8	Ave.	Elevation Sand.	Height Water.	Ver.	Cu. Ft.	Pounds.	Total.	Effective.	Ver.	Sq. Ft.	Weight pr. Sq. Ft.	Friction pr. Sq. Ft.	Date.
11.	10.	7.	10.	9.	9.	7.	6.	8.6	93.	101.6	40.1	39,826	2,485,142	4,445,000	1,959,858	31.5	4989.6	392.6	252.	Feb. 4.
11.	9.	6.	9.	9.	9.	6.	6.	8.1	93.1	101.2	40.2	39,948	2,492,755	....	....	32.1	5084.6	......	.....	" 5.
10.	9.	7.	9.	9.	9.	6.	5.5	8.1	93.	101.1	40.2	39,948	2,492,755	....	....	32.1	5084.6	......	.....	" 6.
....	....	....	....	....	....	....	....	....	....	101.4	....									" 6.
10	9.	6.	9.	9.	9.	6.	6.	8.1	93.4	101.5	41.	40,924	2,553,658	....	....	32.9	5211.4	......	.....	" 8.
11	10.	7.5	10.	9.	9.	6.	6.	8.6	93.5	102.1	42.2	42,388	2,645,011	....	....	33.6	5322.2	......	....	" 9.
12.	11.	8.	9.5	9.	10.5	6.5	7.	9.2	93.5	102.7	43.6	44,096	2,751,590	....	....	34.4	5449.	......	275.2	" 12.
15.	15.	12.	12.	9.	11.	9.	12.	11.9	93.6	105.5	....	....	......			....	......			" 15.
16.	15.5	16.5	15.	15.	12.	11.	12.	14.1	92.6	106.7	....	....	......			....	......			" 17.
18.	18.	18.	20.	26.	24.	12.	15.	18.9	87.3	106.2	....	....	......			....	......			" 19.
16.	20.	17.	19.5	25	27.	12.	16.	19.1	86.8	105.9	47.1	48,244	3,110,426	5,625,000	2,514,574	28.	4435.2	567.	224.	" 20.
27.	22.5	15.	18.	26	27.	17.	15.	21.2	83.9	105.1	....	....	......			....	......			" 22.
....	....	....	....	....	....	....	....	....	....	104.3										" 23.
....	....	....	....	....	....	....	....	....	85.	103.	44.6	45,316	2,827,718	....	....	26.6	4213.4	......	.....	" 24.
15.	....	....	21.	19.5	9.	12.	17.		85.6	102.6	44.5	45,194	2,820,106	5,575,000	2,754,894	27.5	4356.	632.4	220.	" 25.
....									85.5	102.6	44.8	45,560	2,842,944	....	....	27.7	4387.7	......	.....	" 26.
....									85.1	102.6	44.9	45,682	2,850,557	....	....	27.4	4340.2	......	.....	" 26.
18.	25.5	13.	18.	23.	20.	10.	14.	17.7	85.1	102.8	45.1	45,926	2,865,782	5,575,000	2,709,218	27.4	4340.2	624.2	219.2	" 27.
17.	25.	14.	....	22.	19.	13.	13.	17.6	85.1	102.7	45.2	46,048	2,873,395	....	....	27.6	4371.8	......	.....	Mar. 1.
18.	26.	12.	16.	23.	19.	10.	14.	17.3	85.7	103.	45.6	46,536	2,903,846	....	....	28.3	4482.7	......	.....	" 2.
19.	27.	16.	18.	23.	21.	11.	15.	18.8	85.4	104.2	46.9	48,122	3,002,813	....	....	28.1	4451.	......	.....	" 3.
19.	27.	16	18.	23.	21.	11.	15.	18.8	85.	103.8	46.5	47,634	2,972,362	....	....	27.7	4387.7	......	.....	" 4.
....									85.	103.6	46.5	47,634	2,972,362	....	....	27.9	4419.4	......	.....	" 5.
....									85.	102.7	45.7	46,658	2,911,459	....	....	28.	4435.2	......	.....	" 6.
....									85.2	102.3	45.4	46,292	2,888,621	....	....	28.3	4482.7	......	.....	" 8.
....									85.2	102.	45.3	46,170	2,881,008	....	....	28.5	4514.4	......	.....	" 9.
19.	25.	14.	15.	20.	16.	9.	15.	16.6	85.4	102	45.4	46,292	2,888,621	5,585,000	2,696,379	27.8	4403.5	612.3	222.4	" 10.
....	....	....	....	....	....	....	....	....	....	......	....	......	......	....	....	....	......	......	.....	Divers} work.}
....	....	....	....	....	....	....	....	....	....	......	....	......	......	....	....	....	......	......	.....	Total.

APPENDIX F.

Tables showing the Strains in the Fixed Spans, in Pounds.

248 Ft. SPAN.

LENGTH, 246 FT.; END HEIGHT, 22 FT.; CENTRAL HEIGHT, 31.25 FT.; DEAD LOAD ON EACH TRUSS, PER FOOT, 1,330 LBS.; LIVE LOAD, 1,120 LBS.

No. PANEL.	UPPER CHORD.	LOWER CHORD.	MAIN BRACE.	MAIN TIE.	COUNTER BRACE.	COUNTER TIE.
1	86,600	71,514	164,400	172,200	.....	.....
2	225,750	201,363	146,700	153,700	.....	.....
3	350,600	304,993	129,200	142,400	.....	.....
4	417,500	389,445	118,075	123,800	.....	.....
5	482,575	454,115	98,120	113,450	.....	.....
6	530,000	506,385	86,110	95,200	.....	24,860
7	568,000	565,400	65,800	84,410	23,900	24,900
8	588,050	572,560	51,100	65,650	24,650	33,400
9	593,000	583,973	34,300	50,860	33,320	34,300

198 Ft. SPAN.

LENGTH, 198 FT.; END HEIGHT, 22 FT.; CENTRAL HEIGHT, 26 FT.; DEAD LOAD ON EACH TRUSS, PER FOOT, 1,250 LBS.; LIVE LOAD, 1,120 LBS.

No. PANEL.	UPPER CHORD.	LOWER CHORD.	MAIN BRACE.	MAIN TIE.	COUNTER BRACE.	COUNTER TIE.
1	66,700	55,570	115,570	133,936	.....	.....
2	177,500	167,241	106,102	116,490	.....	.....
3	267,200	257,818	88,097	109,095	.....	.....
4	337,500	327,823	77,395	88,884	.....	.....
5	390,200	380,008	58,816	78,470	17,255	17,926
6	424,050	413,592	47,458	59,359	18,091	26,818
7	437,720	428,015	27,654	47,458	26,818	27,654

APPENDIX.

APPENDIX F—*Continued.*

176 Ft. SPAN.

LENGTH, 173.25 FT.; HEIGHT, 22 FT.; DEAD LOAD ON EACH TRUSS, PER PANEL, 14,850 LBS.; LIVE LOAD, 13,860 LBS.

No. PANEL.	UPPER CHORD.	LOWER CHORD.	MAIN BRACE.	MAIN TIE.	COUNTER BRACE.	COUNTER TIE.
1	56,523	48,448	98,816	115,285	.....	.....
2	153,419	145,344	90,276	98,816	.....	.....
3	234,165	226,092	73,828	90,276	.....	.....
4	298,765	290,690	65,310	73,828	.....	.....
5	347,215	339,138	48,840	65,310	15,900	15,900
6	379,511	371,437	40,321	48,840	15,900	23,852
7	395,659	387,585	23,852	40,321	23,852	23,852

130 Ft. SPAN.

LENGTH, 128.3 FT.; HEIGHT, 22 FT.; DEAD LOAD ON EACH TRUSS, PER PANEL, 13,150 LBS.; LIVE LOAD, 16,000 LBS.

No. PANEL.	UPPER CHORD.	LOWER CHORD.	MAIN BRACE.	MAIN TIE.	COUNTER BRACE.	COUNTER TIE.
1	42,510	34,008	67,493	84,366	.....	.....
2	110,526	102,025	60,882	67,493	.....	.....
3	161,539	153,038	43,009	60,882	.....	9,261
4	195,548	187,046	35,397	43,009	9,261	18,523
5	212,552	204,050	18,523	35,397	18,523	18,523

APPENDIX G.

STRAINS IN THE DRAW.

LENGTH OF ONE ARM, 182 FT.; CENTRAL HEIGHT, 34.3; END HEIGHT, 20.8; DEAD LOAD PER FOOT ON EACH TRUSS, 960 LBS.; LIVE LOAD, 1,120 LBS.

For Numbering and Lettering of Ties and Posts, see Plate XII.

Designations of Post.	Quantities taken from Diagram. Moments.	Quantities taken from Diagram. Chord Strains.	Maximum Strains in the Chords. Upper Chord.	Maximum Strains in the Chords. Lower Chord.	Draw Open. No. Tie.	Draw Open. Weight Carried.	Draw Open. Strain in Tie.	Both Arms Loaded. No. Tie.	Both Arms Loaded. Value of z.	Both Arms Loaded. Weight Carried.	Both Arms Loaded. Strain in Tie.	One Arm Loaded. No. Tie.	One Arm Loaded. Value of z.	One Arm Loaded. Weight Carried.	One Arm Loaded. Strain in Tie.
									FORMULA: $S = 69,720 - 224z^2 - 490z$.				FORMULA: $S = 28,620 - 2.887z^2 - 480z$.		
A	20,536,880	598,800		599,080	1	87,360	98,100	1	290.5	151,060	166,170				
B	16,164,512	472,700	363,500	473,080	2	79,920	106,590	2	275	135,170	180,290				
C	12,323,128	363,500	267,000	368,500	3	72,600	98,680	3	259.75	119,990	159,090				
D	8,943,480	267,000	215,500	267,000	4	65,400	87,290	4	244.75	105,500	140,670				
E	7,088,536	215,500	169,000	215,500	5	58,320	77,760	5	230.	91,920	122,560				
F	5,492,336	169,200	132,000	168,000	6	51,360	68,480	6	215.5	78,480	104,640				
G	4,107,376	132,100	99,000	132,000	7	44,520	56,980	7	201.25	65,920	87,890				
H	2,981,160	99,400	78,000	99,000	8	37,800	56,400	8	187.25	53,960	71,930				
I	2,252,432	78,200	92,000	78,900	9	31,200	41,600	9	173.5	42,570	56,760	20	33.	Negative.	6,970
J	2,517,424	91,900	92,000	92,000	10	24,720	32,980	10	160.	31,760	42,350	19	46.5	4,980	21,500
K	2,384,928	92,100	92,000	78,000	11	18,960	24,480	11	146.75	21,480	28,640	18	59.75	15,360	36,820
L	1,854,944	76,300	92,000	76,000	12	12,120	16,160	12	133.75	11,720	15,680	17	72.75	28,300	53,770
M	1,059,968	46,900	92,000	47,000	13	6,600	8,800	13	121	2,470	3,290	16	85.5	38,410	57,330
N			76,000		14	6,000	6,000	14	108.5	Negative.		15	98.	50,980	

Upper Chord grouping: Tension (upper rows), Compression (lower rows).
Lower Chord grouping: Compression (upper rows), Tension (lower rows).

Engineered Irony

APPENDIX H.

List of Persons Employed on the Kansas City Bridge

NAMES.	OCCUPATION.	TIME.
O. Chanute.............	Chief Engineer......................	February 1, 1867, to completion of work.
B. D. Frost...............	Principal Assistant Engineer......	February 1, 1867, to June 18, 1867.
R. H. Temple............	" " "............	June 10, 1867, to February 10, 1868.
W. C. Cranmer..........	Assistant Engineer................	Dec 15, 1866, to June 15, 1867.
C. H. Knickerbocker...	" "............	January 1, 1867, to completion of work.
George Morison........	" "..................	October 15, 1867, " "
William Reineke.......	Draughtsman.....................	April 1, 1867, " "
Joseph Tomlinson.......	Superintendent of Superstructure	October 1, 1867, " "
E. L. Bostwick...........	Superintendent of Carpenters.....	February 1, 1867, " "
John McCollum.........	Foreman of Carpenters............	January 24, 1867, to May 7, 1867.
W. B. Spence............	" "	April 1, 1867, to completion of work.
J. H. Herring.............	" "	April 1, 1867, to January 8, 1868.
P. S. Gidley..............	" "	April 1, 1867, to completion of work.
Edward Kyle............	" "	June 1, 1867, to March 31, 1868.
Joseph Robb............	" "	June 1, 1867, to January 20, 1868.
J. M. Green..............	" "	Nov 1, 1867, to April 4, 1868.
M. W. Vanorman........	" "	April 1, 1868, to November 8, 1868.
E. Davis..................	" "	Nov 13, 1868, to completion of work.
Charles Hutchins........	" "	Sept 1, 1868, to May 8, 1869.
P. McAnany..............	" "	Dec 16, 1868, to May 8, 1869.
W. K. McComas..........	Superintendent of Masonry.......	April 15, 1867, to completion of work.
S. P. Thompson.........	Foreman of Laborers..............	March 29, 1867, to November 31, 1868.
Peter McGee.............	Foreman of Pile Drivers...........	October 31, 1868, to May 15, 1869.
John McGee.............	Foreman of Boatmen..............	October 1, 1867, to completion of work.
W. H. Crampton.........	Superintendent of Laborers.......	February 20, 1867, to July 10, 1868.
Joseph Neville...........	Foreman of Laborers..............	February 1, 1867, to June 7, 1867.
A. J. Crouse..............	" "	June 10, 1867, to July 20, 1867.
R. S. Scott...............	" "	October 1, 1867, to January 12, 1868.
V. C. Wood..............	" "	June 1, 1867, to completion of work.
Lyman Beebe............	Master Blacksmith.................	April 15, 1868, to May 15, 1868.
Joseph Thayer............	" "	June 1, 1868, to completion of work.

APPENDIX.

APPENDIX H—*Continued.*

NAMES.	OCCUPATION.	TERM.
Peter Scully............	Submarine Diver...............	September 1, 1867, to September 30, 1868.
G. A. Bailey............	" "	September 13, 1868, to January 5, 1869.
Joseph Battles...........	" "	February 15, 1869, to June 6, 1869.
E. P. Harrington........	" "	February 16, 1869, to March 15, 1869.
J. H. Phillips...........	" "	February 16, 1869, to March 15, 1869.
W. C. Perry.............	" "	February 16, 1869, to March 15, 1869.
J. S. Quinn.............	" "	February 16, 1869, to March 15, 1869.
J. W. Van Norman.......	" "	February 16, 1869, to March 15, 1869.
J. H. Cowing...........	" "	February 16, 1 69, to March 15, 1869.
Moses Torrance..........	" "	December 14, 1868, to March 7, 1869.
C. Ryan................	Captain Str. "Gypsey".........	January 1, 1867, to October 16, 1867.
J. S. Bellas.............	" "	October 20, 1867, to completion of work.
J. A. Wise.............	Pilot Str. "Gypsey"...........	June 5, 1867, to February 18, 1868.
T. J. Boone............	" "	February 15, 1868, to April 20, 1868.
J. N. Montgomery.......	" "	May 26, 1868, to completion of work.
Thomas Newkirk........	Engineer Str. "Gypsey"........	June 10, 1867, to October 17, 1867.
D. C. Riter............	" " "	October 13, 1867, to November 21, 1867.
A. W. Hardy...........	" " "	December 1, 1867, to completion of work.
J. R. Balis.............	Auditor......................	February 20, 1867, to completion of work.
G. E. Pitkin............	Accountant...................	July 15, 1867, " "
Employed by Masonry Contractors :		
Nelson Gautier...........	Superintendent of Masonry......	September 19, 1867, to completion of work.
Adrian Mitchel...........	Foreman of Stone-Cutters.......	September 19, 1867, " "
Maurice Scanlon.........	" " "	March 4, 1867, to December 20, 1867.
Peter Shipner...........	" " "	December 20, 1867, to August 26, 1868.
P. K. Smith.............	Bookkeeper...................	March 1, 1868, to completion of work.
Employed by Keystone Bridge Co. :		
Clement McMahon........	Foreman of Raisers on Draw.....	
H. M. Shotts............	Asst. Foreman of Raisers on Draw	
F. S. Kauffman..........	Foreman of Raisers on Fixed Spans	
T. J. Bell.............	Foreman of Framers on Fixed Spans	

ERRATA.

Page 14, line 10, for equivalent read equipment.

  "   23, "   22, " inpinging   "   impinging.

  "   31, "   31, " gates   "   gate.

  "   40, "   31, " work   "   rock.

  "   42, "   20, " bottom   "   batter.

  "   48, "   8, " An   "   A.

  "   51, "   11, " the pier   "   this pier.

  "   53, "   31, " 20   "   22.

  "   54, "   21, " basis   "   bases.

  "   56, "   20, " end   "   rod.

  "   65, "   13, " walls   "   wells.

  "   85, "   18, " and   "   area.

  "   86, "   6, " support   "   supports.

  "   89, "   29, " stood   "   chord.

  "   95, "   4, " land   "   hand.

  " 105, the lines A B and A D, in Fig. 3, should be parallel with the corresponding lines in Fig. 4.

  " 106, line 16, for $\frac{4 \times 2.5}{6.15}$ read $\frac{4 \times 2.5}{61.5}$.

  " 106, " 23, " $\frac{e}{243}$ " $\frac{e}{246}$.

  " 107, " 22, " $\frac{e}{243}$ " $\frac{e}{246}$.

  " 109, " 9, read $- M \div 34.3 = 463500$ pounds.

  " 112, " 17, " $\frac{d\,m}{d\,x} = 1660\,l - 2080\,x$.

  " 112, " 28, " $w = 480$, $w' = 560$ and $l = 290.5$.

Engineered Irony

ILLUSTRATIONS.

PLATES.

I.—MAP SHOWING LOCATION OF BRIDGE.

II.—WATER RECORD—CROSS SECTION OF RIVER—PROFILE OF CROSSING—PONTOON PRO-
TECTION.

III.—WATER DEADENER—CAISSON No. 2—FOUNDATION WORKS, PIER No. 3.

IV.—FOUNDATION WORKS, PIER No. 4.

V.—FOUNDATION WORKS, PIER No. 4.

VI.—CAISSON No. 5—SHEET PILING AT PIER No. 6—DETAILS OF DREDGES—PILE SHOE—
BETON BOX.

VII.—MASONRY—DRAW PROTECTION—FALSE WORKS BETWEEN PIERS 3 AND 4.

VIII.—FLOATING DERRICKS.

IX.—GENERAL ELEVATION—176 FEET SPAN.

X.—248 FEET SPAN.

XI.—PLANS OF DRAW.

XII.—STRAIN DIAGRAMS.

Engineered Irony

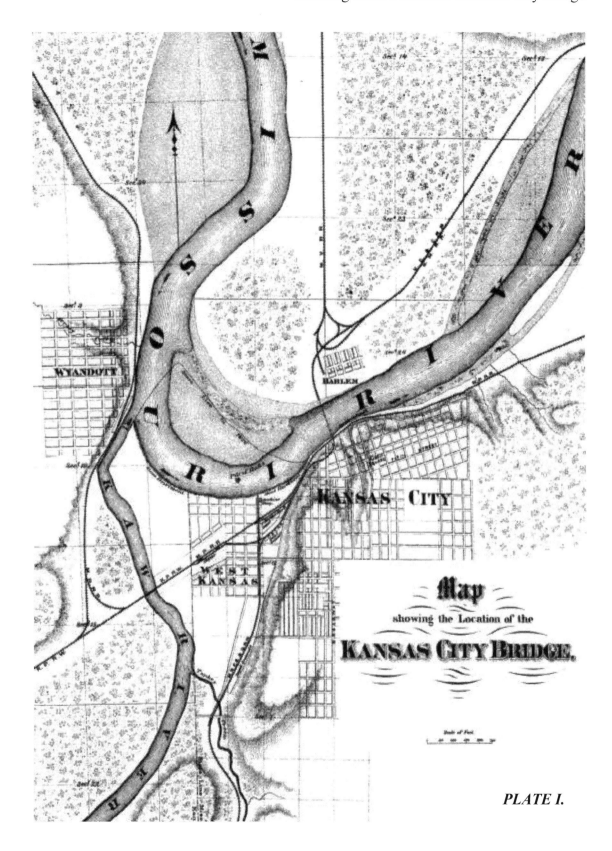

PLATE I.

Engineered Irony

Crossing Octave Chanute's Kansas City Bridge

Engineered Irony

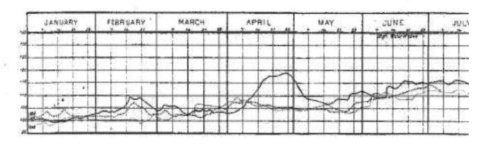

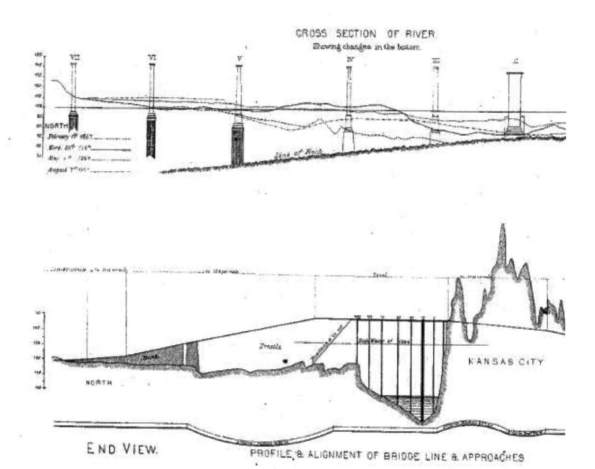

PLATE II.

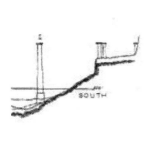

SOUTH

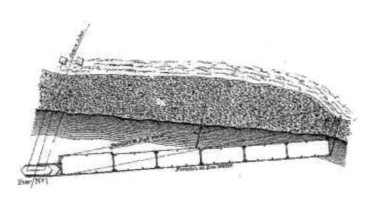

PLAN OF PONTOON PROTECTION
For Steamboats passing down through Draw.

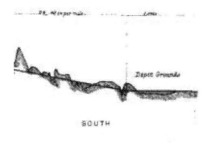

SOUTH

Scale of Feet.

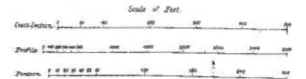

D.Van Nostrand, Publisher, N.Y.

Engineered Irony

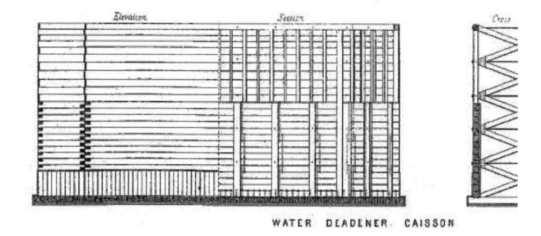

WATER DEADENER CAISSON

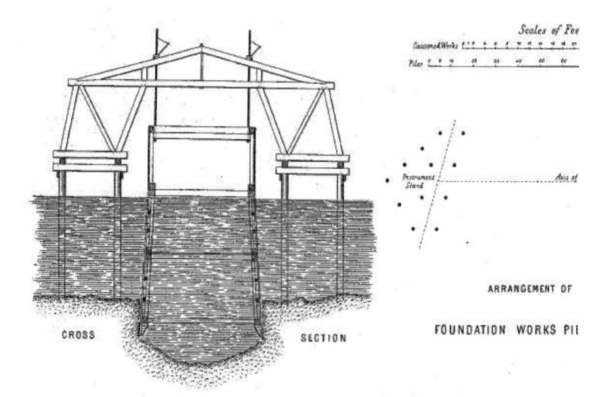

Scales of Fee

Caisson Works

Pier

CROSS SECTION

ARRANGEMENT OF

FOUNDATION WORKS PI

PLATE III.

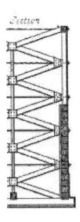

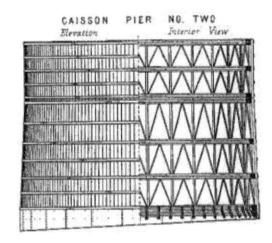

CAISSON PIER NO. TWO
Elevation   *Interior View*

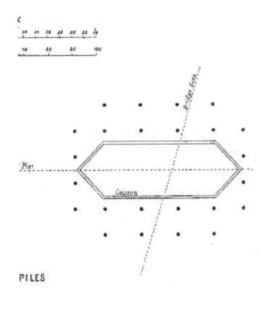

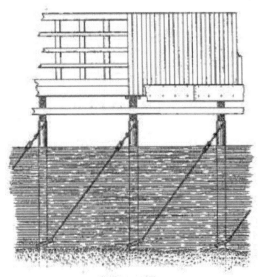

PILES

R NO. THREE.

SIDE VIEW
Showing Caisson partly built
A method of securing the Piles

D. Van Nostrand, Publisher, N.Y.

Engineered Irony

Engineered Irony

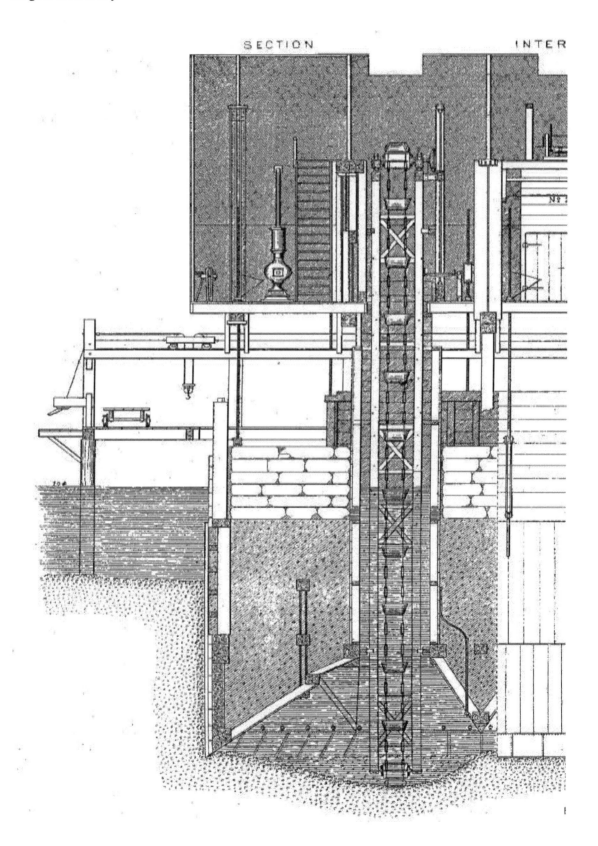

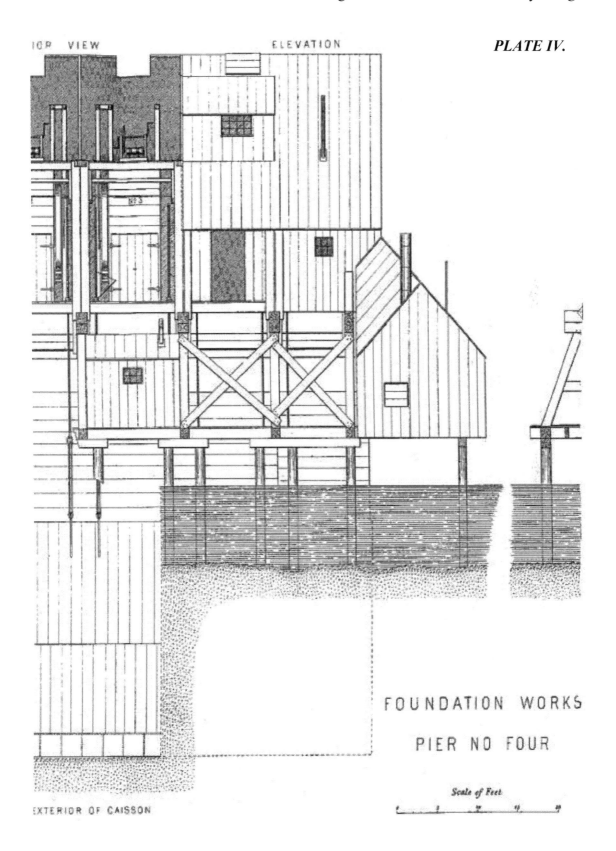

PLATE IV.

ICR VIEW

ELEVATION

FOUNDATION WORKS

PIER NO FOUR

EXTERIOR OF CAISSON

Scale of Feet

Engineered Irony

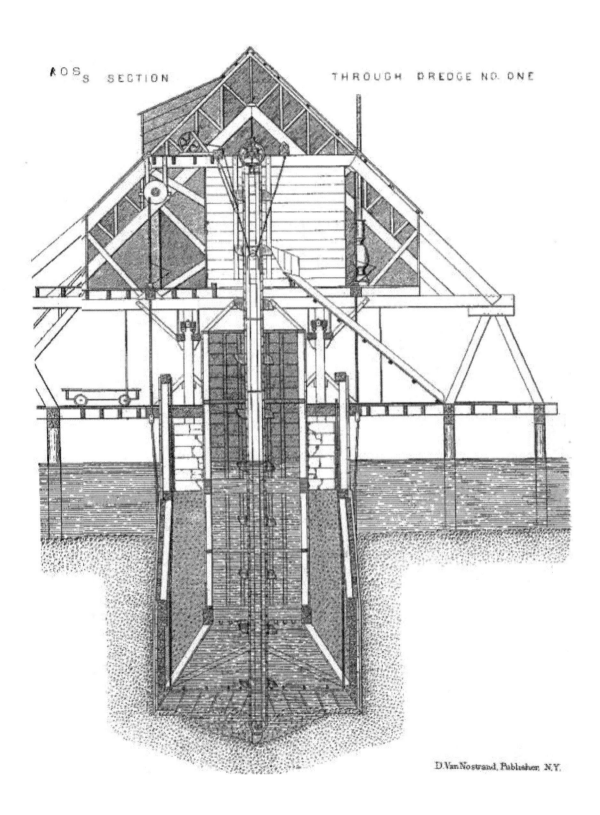

ROSS SECTION   THROUGH DREDGE NO. ONE

D. Van Nostrand, Publisher, N.Y.

Engineered Irony

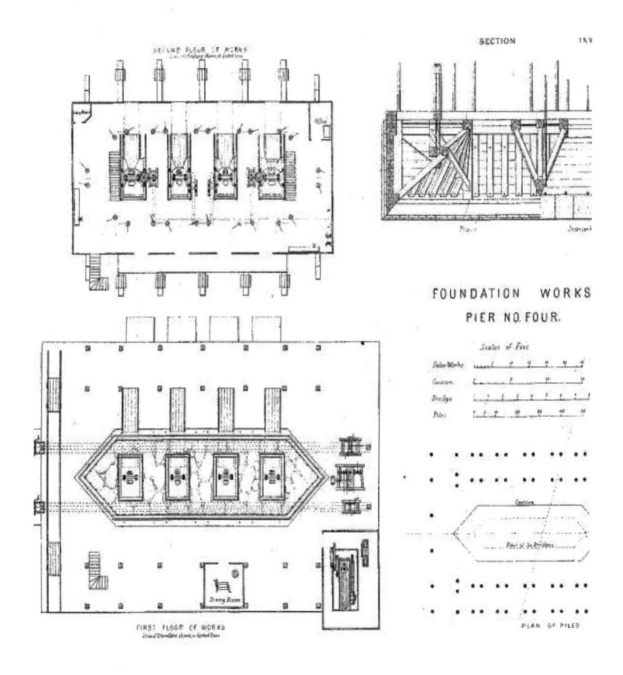

PLATE V.

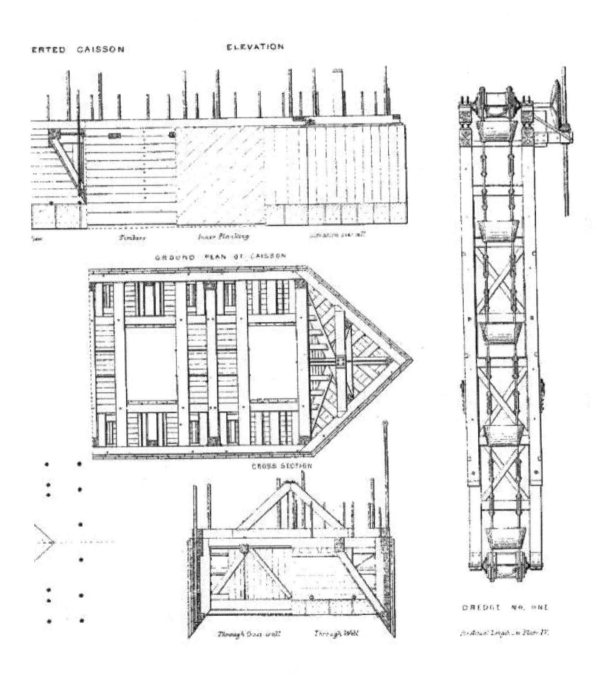

Engineered Irony

Crossing Octave Chanute's Kansas City Bridge

Engineered Irony

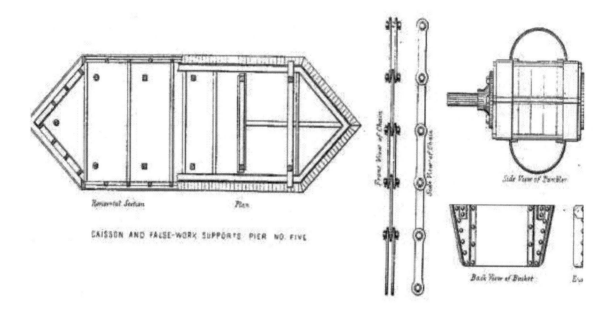

Horizontal Section   Plan

CAISSON AND FALSE-WORK SUPPORTS. PIER NO. FIVE

Front View of Chain   Side View of Chain

Side View of Tumbler

Back View of Bucket   End

DETAILS OF DREDGE N

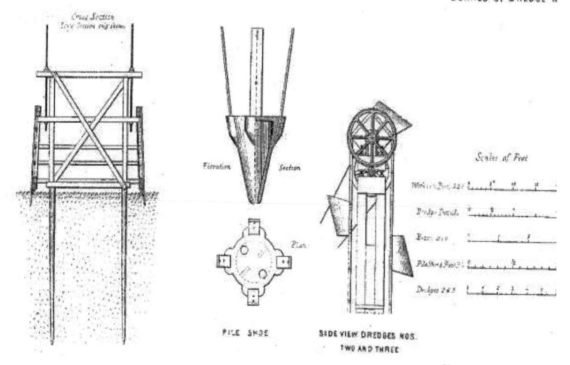

Cross Section
Loose frame coffer dam

Elevation   Section

Plan

PILE SHOE

SIDE VIEW DREDGES NOS.
TWO AND THREE

Scale of Feet

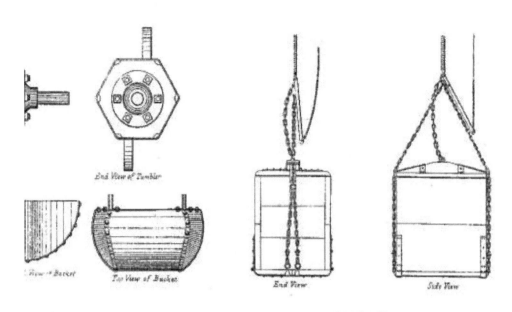

End View of Tumbler

View of Bucket

Top View of Bucket

End View

Side View

O ONE

BETON BOX

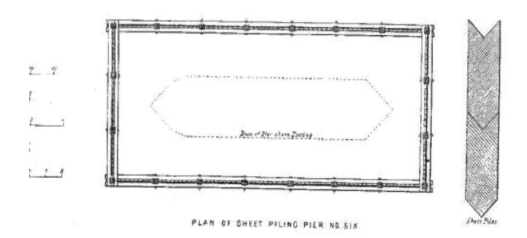

PLAN OF SHEET PILING PIER NO. SIX

Sheet Piles

MAYER & MERKEL, 111 FULTON ST. NEW YORK
D.Van Nostrand, Publisher, N.Y.

Engineered Irony

MASON

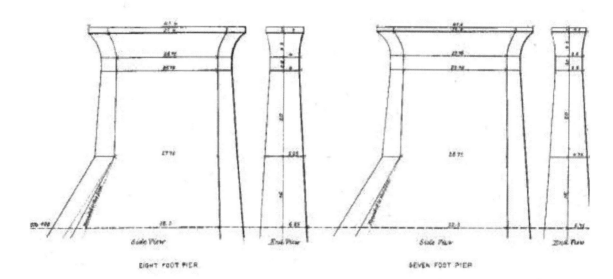

EIGHT FOOT PIER   SEVEN FOOT PIER

TEMPORARY DRAW PROTECTION.

(Used as False-Works for raising the Iron Superstructure.)

Masonry

Falsework

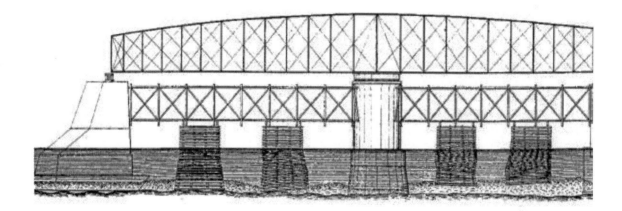

R Y

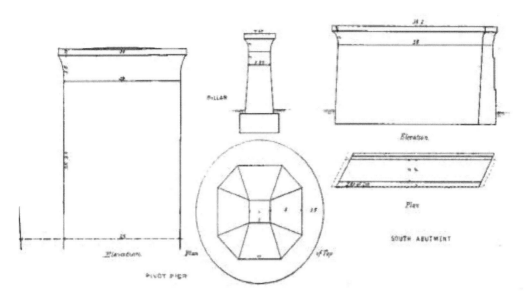

PILLAR

Elevation.

Plan.

South Abutment

Elevation.

Plan.

of Top.

PIVOT PIER

Scale of Feet.

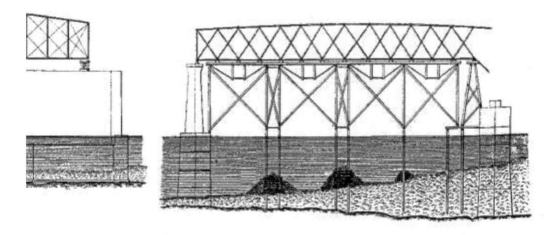

FALSE WORKS USED IN RAISING SUPERSTRUCTURE

BETWEEN PIERS 3 & 4

MAYER & MERKEL, 91 FULTON ST. NEW YORK

D. Van Nostrand, Publisher. N.Y.

Engineered Irony

Crossing Octave Chanute's Kansas City Bridge

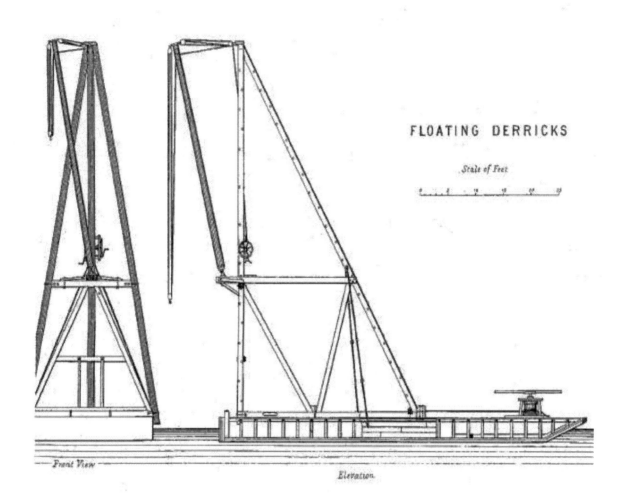

FLOATING DERRICKS

Scale of Feet

Front View

Elevation

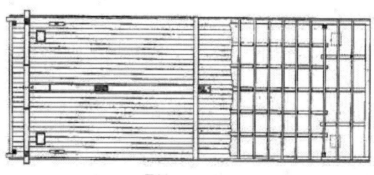

Plan

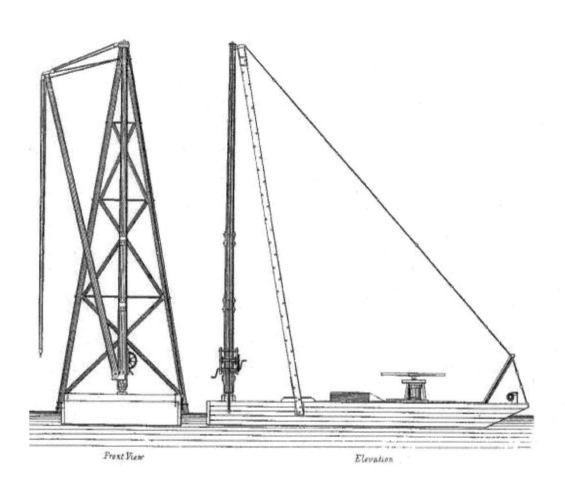

Front View   Elevation

NO. TWO
(Smaller Boat)

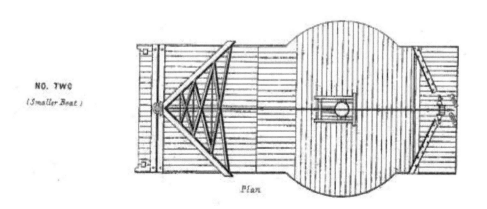

Plan

D.Van Nostrand, Publisher, N.Y.

Engineered Irony

Crossing Octave Chanute's Kansas City Bridge

Engineered Irony

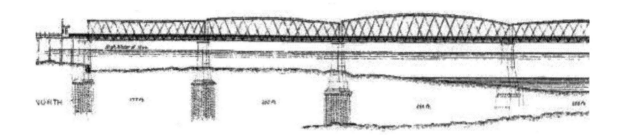

NORTH

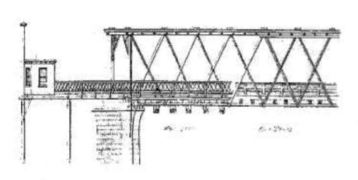

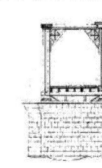

ELEVATION

END VIEW

PLATE IX.

E W.

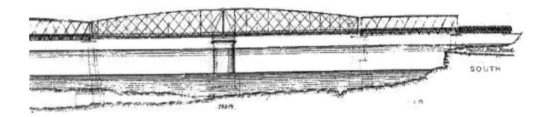

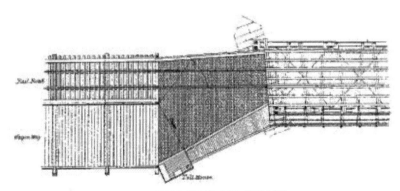

PAN

PLAN SHOWING APPROACH.

SOUTH

D.VanNostrand, Publisher, N.Y.

Engineered Irony

Crossing Octave Chanute's Kansas City Bridge

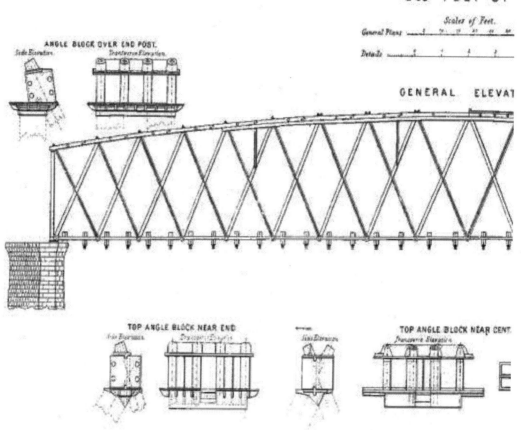

248 FEET SP

Scales of Feet.

General Plans

Details

GENERAL ELEVAT

ANGLE BLOCK OVER END POST.

TOP ANGLE BLOCK NEAR END

TOP ANGLE BLOCK NEAR CENT

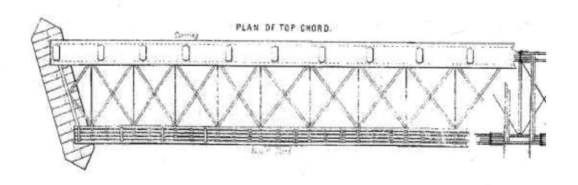

PLAN OF TOP CHORD.

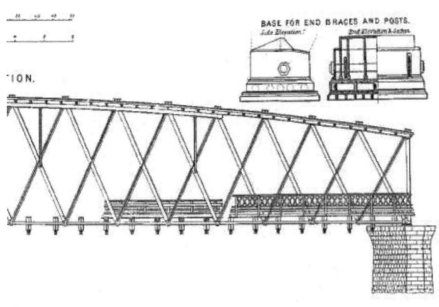

PLATE X.

A N.

·ION.

BASE FOR END BRACES AND POSTS.

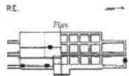

RE.

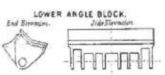

LOWER ANGLE BLOCK.

PLAN OF BOTTOM CHORD.

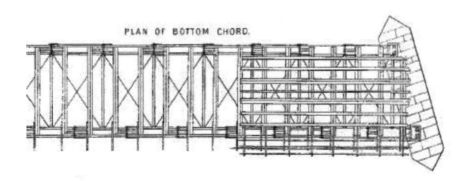

MAYER & MERKEL, 81 FULTON ST. NEW YORK.
D.Van Nostrand,Publisher, N.Y.

Engineered Irony

Crossing Octave Chanute's Kansas City Bridge

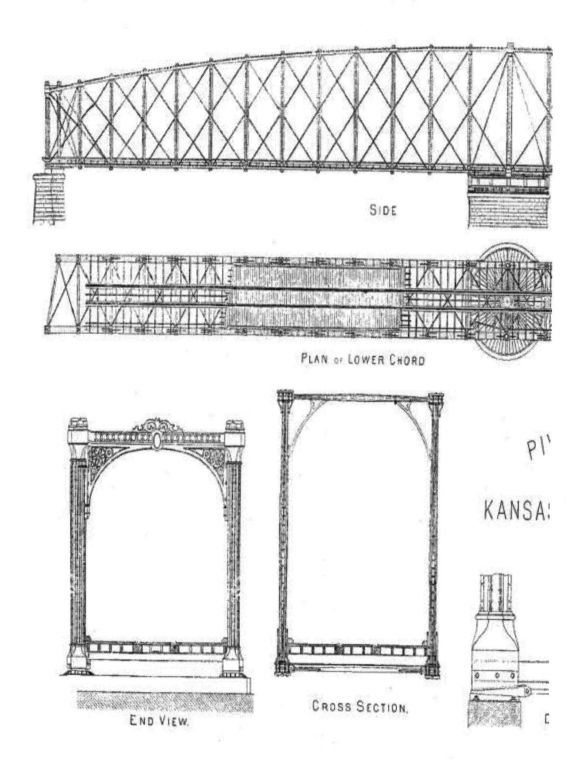

SIDE

PLAN of LOWER CHORD

PI'

KANSA:

END VIEW.

CROSS SECTION.

PLATE XI.

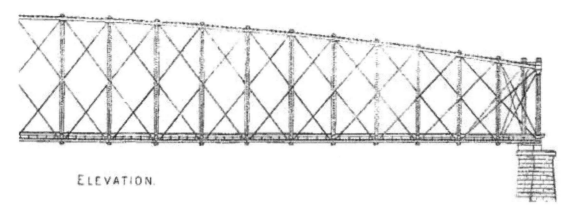

ELEVATION.

PLAN of UPPER CHORD.

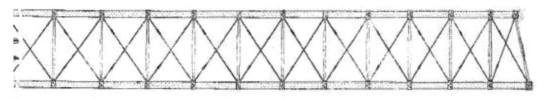

JOT SPAN

OF THE

S CITY BRIDGE.

LINVILLE AND PIPER PATENT

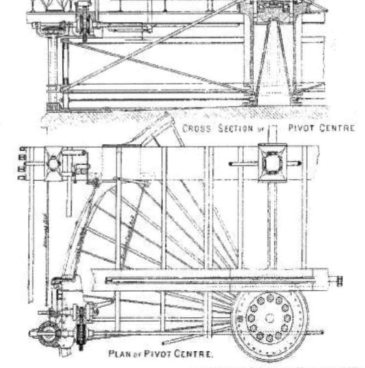

CROSS SECTION of PIVOT CENTRE.

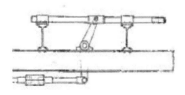

DETAIL of WEDGES.

PLAN of PIVOT CENTRE.

MAYER & MERKEL. 91 FULTON ST. NEW YORK.

Engineered Irony

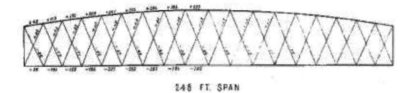

248 FT. SPAN

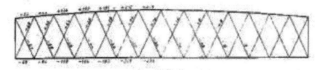

198 FT. SPAN

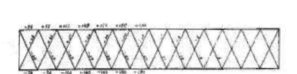

176 FT. SPAN

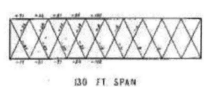

130 FT. SPAN

STRAIN SHEET

Scale of Feet

Scale of Strains Pound

Figures on Indicate Diagrams Area Observations Strains in First Compression or Extension

Proportion of Posts & Ties

DRAW

PLATE XII.

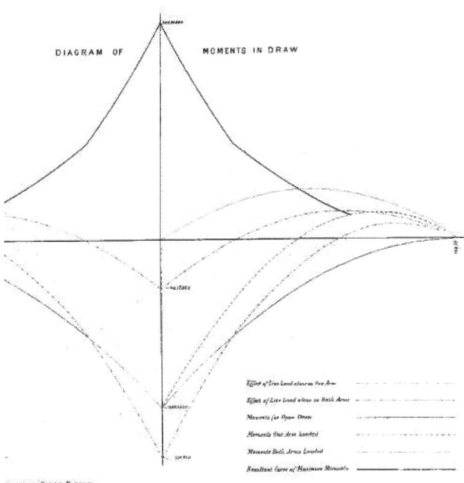

DIAGRAM OF   MOMENTS IN DRAW

NVILLE and PIPER PATENT.

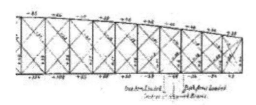

Engineered Irony

OCTAVE CHANUTE.
"Father of Aviation."

February 18, 1832   –   November 23, 1910
Left Bank, Seine River, Paris, France   Chicago, Illinois

"Chanute's nature was mild, considerate to others, kindly, generous, patient, and just. Yet while almost self-effacing in his manner, it was recognized by all with whom he came in contact that he was a tremendous thinker and doer, with intense tenacity of purpose and determination." --*Simine Short*

by Bill Nicks, Jr.
with contributions by David W. Jackson

It's not hard to imagine a young, inquisitive, 6-year-old boy named Octave Alexandre Chanut—having recently immigrated to New Orleans, Louisiana, from Paris, France, in fall 1838—sitting around a dining room table with his father, Joseph Chanut, and other Jefferson College professors listening to discussions on interesting topics of the day.

Nor is it hard to imagine that as an adult, the same Octave Chanute (he added the 'e' at age 21) would lead similar discussions (not across a dining room table; but, in some instances, across continents) on bridge building, rapid transit, wood preservation, and aviation.

Octave Chanute, ca. 1856.

Image courtesy Simine Short, author of *Locomotive to Aeromotive: Octave Chanute and the Transportation Revolution*, as photographed in 2010 from the Octave Chanute Papers PR 13 CN 1973: 232 Container 1, Folder1, Manuscript Division, Library of Congress, Washington, D.C.; digitally enhanced by *Engineered Irony* author David W. Jackson.

Octave Alexandre Chanut was born on Left Bank of the Seine River on February 18, 1832, in Paris, France. His father, Joseph Chanut, accepted a position as Vice President and History Professor at Jefferson College, north of New Orleans, Louisiana, in fall 1838. His son joined the immigration to the New World.[8]

In 1846, Chanut and his father moved to New York City. The month-long steamship voyage along the Mississippi and Ohio Rivers through the state-owned canal-railroad system across Pennsylvania to New York left a lasting impression on the youth. He was fascinated with modern technology and engineering.

While Joseph Chanut was engaged in literary pursuits, the young Chanut received his education in a New York boarding school. Engineers earned their "C.E." (Civil Engineer) in the field. In 1848, the teenager applied for a job with the Hudson River Railroad. When his application was turned down, he offered to work for free. A few weeks later he was hired as a chainman, the lowest paid position at the railroad. His future co-collaborator on the Kansas City Bridge, George S. Morison, also started out in the engineering profession as a chainman.

The 1850 U.S. Census shows Chanute staying at the Marshall Tavern in Hyde Park, New York.

In 1854, he became a naturalized American citizen. He Americanized his name by adding the letter "e" to his surname, and dropping his middle name. He became Octave Chanute.

Chanute launched a long and distinguished career as a civil engineer in the transportation field. He worked for many of the major railroad companies, always looking

to improve systems, always generating fresh ideas. He led discussions on, and improvements to, such topics ultimately landed him in numerous scientific Halls of Fame. However, prior to those honors he had to learn and practice the profession of civil engineering.

Prior to moving to Kansas City in 1867 to build the Kansas City Bridge for the Hannibal and St. Joseph Railroad Chanute married on March 12, 1857, to Annie Riddell James of Peoria, Illinois (they had six children).

Annie Riddle James Chanute, 1860s.

Image courtesy Simine Short, as photographed from the Octave Chanute Papers, Manuscript Division, Library of Congress, Washington, D.C.; digitally enhanced by David W. Jackson.

Chanute held various professional positions, including:

1. Worked on the Chicago & Alton RR (1853);

2. Directed construction of the longest RR drawbridge in the world over the Illinois River at Peoria, Illinois (1856);

3. Directed the maintenance on the Chanute Surveyed and laid out the line from Peoria to the Indiana border (112 miles) for the Toledo, Peoria & Warsaw RR; platted Fairbury, Livingston County, Illinois (1857);

4. Served as Division Engineer of Maintenance of Way of the Pittsburgh, Fort Wayne & Chicago RR (1861);

5. Chief Engineer of the Ohio & Mississippi RR (1862)

6. Chief Engineer of the Chicago & Alton RR (1863); and,

7. Won the designed challenge and oversaw construction of the Union Stockyards in Chicago, Illinois (1863-65).

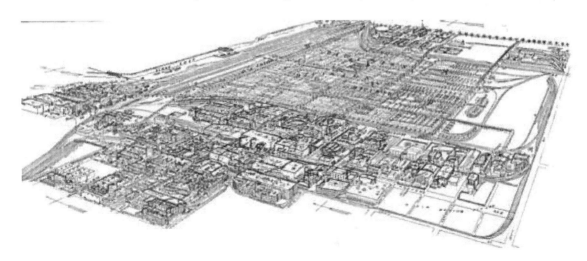

"Rascher's birds eye view of the Chicago packing houses & Union Stock Yards" Rascher, Charles. 1890. Courtesy Library of Congress, Geography and Map Division.

The Chanutes lived in Kansas City for a number of years spanning either side of the Kansas City Bridge project. On February 28, 1867, the *Kansas City Daily Journal of Commerce* reported that Chanute "started for Chicago to bring his family on and to make purchases for the bridge."[9]

Chanute's 1867 move to Kansas City was to bridge the "…rapid, shifting, and ill-reputed" Missouri River. At the time, many thought it was impossible to bridge the mighty river due to the swift and dangerous currents. However, his own knowledge coupled with the advice and experience of the local American Indians, early settlers and bridge-builders on the Rhine River in Germany subdued the river.

"He wanted to learn as much as he could about the 1844 flood and picked the brains of longtime residents to locate highwater marks. His bridge would need to be high enough to avoid damage during a historic flood but not so high that the grade into the West Bottoms became too steep for heavily loaded trains to safely descend. The greatest test of his calculations came with a late spring flood that killed 19 people in Kansas City in 1903. The bridge remained above the waterline and unscathed."[10]

The Chanute family "lived in a stone castle on a hill at Seventh and May Streets."[11] "Chanute amused himself flying curious kites on the bluffs."[12]

**First National Bank
Kansas City, Missouri**

Kansas City Daily Journal of Commerce, April 15, 1873. State Historical Society of Missouri—Kansas City Research Center, K0395, Box 12, F318-320; also, *Campbell's Gazetter*, 1875. Chanute was a director of the bank from at least January 18, 1873, and as late as April 21, 1877.

He and "his engineering staff established their offices over the First National Bank, an 'elegant building on the corner of Delaware and Fourth [sic.]Streets." [actually, Fifth Street[13]]

Chanute invested in real estate and other ventures like the West Bottoms Land Company, and Omnibus and Carriage Company.[14]

"He was once superintendent of the old Kansas City & Southern Railway, which was merged into the Santa Fe system."[15]

While constructing the Kansas City Bridge, from 1867-July 1869, Chanute was also in charge of building cattle and rail lines into Kansas for: the Kansas City, Fort Scott, and Memphis RR, and for the Atchison, Topeka & Santa Fe RR.

During the summer of 1869, being a bit of a land speculator, he purchased from "Squire Charles A. Bradshaw 41.5 acres near the Missouri River, Fort Scott and Gulf Railroad (later Frisco Railroad) right-of-way, and used it to plat the town of Lenexa, Kansas. It is said that the town was originally going to be called Bradshaw, but Bradshaw refused out of modesty. On August 26, 1869 [the month after the Kansas City Bridge was dedicated], Chanute sold the property to three men from Jackson County, Missouri."[16]

In 1871, Chanute was the Kansas City city engineer. He designed the Kansas City Stockyards in the West Bottoms that would reign for more than 100 years before closing and being razed in the early 1990s.[17]

Following the construction of the Kansas City Bridge, Chanute worked for the Leavenworth, Lawrence & Galveston RR (LL&G), building lines south through Kansas. In 1870, his LL&G route took him through Neosho County, Kansas, near the four little towns of Tioga, New Chicago, Chicago Junction, and Alliance. Three years later on January 1, 1873, those four towns merged into one and took his name, becoming Chanute, Kansas.[18]

Chanute's reputation as a problem solver and a quality engineer landed him the chairmanships on two committees for the American Society of Civil Engineers. One committee (1875) was charged with studying mass and rapid transit for New York City. The published report recommended, among other improvements, four elevated lines along the Avenues. The recommendations were accepted and 'The El' was built.

The other committee (1880-85) was to report on wood preservation for railroad ties. Wood preservation directly affected rail safety and maintenance. The finished report recommended the use of chemicals and creosote and was the authority on wood preservation for many years. Later, Chanute introduced the railroad "date nail" to the United States, there making it possible to record the age and viability of preserved railroad ties. In 1885, several railroads asked him to build plants for the preservation of timbers. He and his friend, Joseph P. Card, started the Chicago Tie Preserving Company. This successful business venture allowed him the finances and the time (he retired from Civil Engineering) to take up the question of manned powered flight.

In 1889, he began corresponding with his many contacts from all over the world who were interested in conquering the air.[19] He catalogued the records of past flight experiments. After studying these experiments, he wrote 27 magazine articles on them beginning in October 1891. In these articles he offered his opinions on how the most promising experiments could be improved. The articles appeared in the *American Engineer and Railroad Journal*. In 1894, the articles were republished as his book, *Progress in Flying Machines*.

Two years later, at age 64, Chanute began his own glider experiments near Gary, Indiana, just off Lake Michigan. His nearly 2,000 glider flights made use of a monoplane, then multi-plane gliders, and finally, the more efficient biplane. Through his writings and his experiments, Chanute established himself as an expert in aviation.

The Chanutes continued to venture to Kansas City. In July 1895, Chanute visited his sister and brother-in-law, Mr. and Mrs. C. P. James, who then lived at 2613 Troost Avenue. While in the city, "he viewed the bridge as he passed over it on the cars a few days ago, and he spent considerable time looking about the city. The building at the northwest corner of Fourth and Delaware Streets where the bridge company had its principal offices stands as formerly, but the neighborhood is greatly changed."

Changing Neighborhood: Wyandotte Between 3rd and 4th Streets

One block east from where Octave Chanute once engineered the Hannibal Bridge and directed its construction from the First National Bank building, was the Madam Lovejoy Mansion, northwest corner of 4th and Wyandotte (later operated by Madam Cora Totty). To the left in this photo was Eva Prince's brothel, and to the right was Annie Chambers' "resort" at 201 W. 3rd St. (southwest corner 3rd and Wyandotte).

Late night entertainment at least from the 1970-1990s at this location included adult films and live "strip" shows at The Old Chelsea. Then, in 2005, the neighborhood gentrifying, a sleek, new office building at 300 Wyandotte became the home to HOK Sport of Kansas City, renamed Populous in 2019, which moved to 4800 Main (former Board of Trade Building) in the Kansas City Country Club Plaza District. As of December 2020, under new management, the massive building in the former red light district is leased by a number of business tenants, including the Economic Development Corporation of Kansas City, Missouri.

This was the newspaperman's inconspicuous way of evading the fact that the Chanute's former office spaces and neighborhood had become Kansas City's "red light district." Kansas City's most notable madam, Annie Chambers—one block west at the southwest corner of Third and Wyandotte—operated her "resort" that from outward appearances resembled more of a 'warehouse.' And across the street from Chanute's former quarters at the First National Bank Building, Madam Lovejoy operated an elegant Queen Anne mansion (later operated by Madam Cora Totty). And, next to Lovejoy's was Eva Prince's brothel.

Still, Chanute said, "'The city [is] progressing . . . .' When the bridge was built cable lines were an unknown factor in the city's growth and at that time Mr. Chanute never dreamed of living [way] out on Troost Avenue, where he was today."[20]

On May 30, 1899, Wilbur Wright wrote to the Smithsonian Institution seeking advice on mechanical human flight. The reply contained a list of publications on aerial navigation. Included on the list was Chanute's book, *Progress in Flying Machines*.

A year later, May 13, 1900, Wilbur Wright first wrote to Chanute. Chanute's May 17th response offered assistance, and it began a relationship with the Wright brothers that lasted until Chanute's death, documented in nearly 500 pieces of correspondence between the innovators.[21] Chanute also met with the Wrights in Chicago, Dayton, Ohio, and Kitty Hawk, North Carolina.

In fact, Chanute was at Kitty Hawk in early November 1903, just a few weeks before the Wrights successfully flew. He left Kitty Hawk due to the rough living conditions (he was 71 at the time) and delays caused by damaged propeller axles. However, Chanute was one of the first to know of the Wrights' success. A telegram detailing the first four flights of December 17, 1903, arrived at the Wrights' Dayton home at 5:25 p.m. Chanute, in Chicago, received a telegram from Katherine Wright (the brothers' sister) at 8:07 that evening.

Later that month, Chanute announced to the world, at an American Association for the Advancement of Science meeting, that the Wrights had flown.

Annie Chanute died in 1902. Chanute last visited Chanute, Kansas, in 1906, most likely on a visit to see (or, perhaps dedicate) a new building at 407 W 8th St in Kansas City's historic garment district that was named after the city's former engineer.

Octave Chanute died in Chicago on November 23, 1910, after contracting pneumonia while traveling in Europe with his daughters. He, Annie, and their family rest in peace in the Springdale Cemetery and Mausoleum in Peoria, Illinois.

The Kansas City Star in an obituary titled, "Built the Hannibal Bridge," noted "Kansas City remembers Octave Chanute best for that achievement here."[22]

Earlier that year he wrote the chapter, "Recent Progress in Aviation," for the 1910 *Smithsonian Institution Annual Report*.

The Western Society of Engineers, in May 1911 published an extensive tribute to:

Octave Chanute
In Memoriam

This memoir records the professional career of an Engineer closely identified for the last sixty years with the development of transportation on land and in the air.

Octave Chanute was born in Paris, France, February 18, 1832. His father was Professor of History in the Royal College of France, in Paris, and in the year 1838 accepted the appointment of Vice-President of Jefferson College in the State of Louisiana; he was a resident of Louisiana until 1844, when he removed to New York City and engaged in literary pursuits. His son, Octave, who was six years old when the family came to this country, completed his education in New York, and became, to use his own expression, thoroughly Americanized.

Octave Chanute.
Born Feb. 18, 1832.   Died Nov. 23, 1910.

FROM PHOTO. TAKEN IN FRANCE.

Mr. Chanute began work at the age of seventeen in 1849, on the Hudson River Railroad. This beginning was made, as was usual at that day, at the very foot of the ladder. Young Chanute introduced himself to the Resident Engineer at Sing-Sing and asked for employment. When told that there was no vacancy, he asked for permission to serve without pay as a volunteer chainman. This was somewhat reluctantly granted, with the result that within two months the young man was put on the payroll at $1.12½ per day, and thought his fortune made. Nor was he far wrong, for Mr. Chanute has often said that this was the only position he ever applied for and that he had been continuously engaged since that time without having to solicit employment.

He remained with the Hudson River Railroad for four years, until the completion of construction, when he became Division Engineer at Albany, in charge of terminal facilities and of maintenance of way.

In 1853, immigrants were pouring into Illinois to buy government lands at $1.25 an acre which are now selling at more than $50.00 an acre, and railroad construction was proceeding rapidly. Mr. Chanute came west with Mr. H. A. Gardner, his former Chief Engineer on the Hudson River Railroad, and was engaged on the surveys and construction of what is now a portion of the Chicago & Alton Railroad, between Joliet and Bloomington. Before this was quite completed he became Chief Engineer of the eastern portion of what is now the Toledo, Peoria & Western Railroad, and built that Road from Peoria to the Indiana State Line, a distance of 112 miles. Upon its completion in 1857 he remained in charge of maintenance of way until 1861, when, the Road having gone into the hands of a Receiver, he accepted the position of Division Engineer of Maintenance of Way of the Pittsburgh, Fort Wayne & Chicago Railroad between Chicago and Fort Wayne. This appointment was given him by his old friend and Chief, Mr. Gardner, who one year later recommended him for the position of Chief Engineer of the reconstruction and maintenance of the Western Division of the Ohio and Mississippi Railroad, from St. Louis to Vincennes. This he accepted, but six months later the Road changed hands, by one of those sudden vicissitudes not uncommon in those days, and Mr. Chanute, having by that time attained a reputation for industry, efficiency, and fidelity, received simultaneously several offers of employment. He accepted that of the Railroad which he had first served in the west and became the Chief Engineer of the Chicago & Alton Railroad in 1863, and as such took charge of the reconstruction which, by that time, had been found necessary, for these early American railroads were cheaply built. His duties included the maintenance of way and the building of an extension of the line from Alton to St. Louis.

During this connection, which lasted until 1867, Mr. Chanute submitted a competitive design for the Union Stock Yards of Chicago, and this having been selected in preference to a score of other designs, he was made Chief Engineer of the Yards, and supervised their construction in addition to his railroad duties.

These various engagements brought him in contact with prominent railroad men, and he was next offered the design and construction of the pioneer bridge over the Missouri River at Kansas City. This offer was accepted, and he tendered his resignation as Chief Engineer of the Chicago & Alton Railroad. The Board of Directors of that Railroad, accepting his resignation, passed a resolution of regret at the severance of his connection with that company. The construction of the Kansas City bridge, across a stream so rapid, shifting, and ill-reputed as the Missouri River, involved what were, at that time, a number of novel engineering problems. It was successfully completed in July, 1869, and attracted general interest as the first bridge built over the Missouri River. A book, giving an account of the construction of the Kansas City Bridge, was written by Mr. Chanute and Mr. George S. Morison, his principal assistant Engineer, and published in 1870.

While engaged in completing the Kansas City bridge, Mr. Chanute was placed in charge of the building of several railroad lines, extending into Kansas, planned to secure a portion of the cattle trade coming overland from Texas: first that of the Kansas City, Ft. Scott & Memphis Railroad, from Kansas City to the line of the Indian Territory, then of a parallel line, now known as the Southern Kansas division of the Atchison, Topeka & Santa Fe Railroad, then of a connecting line between these two, and lastly of a line from Atchison northward, the whole comprising the construction of about 400i miles of railroad, and incidentally thereto the design and construction of the Union Stock Yards of Kansas City.

These various works were completed in 1871, and Mr. Chanute then became the General Superintendent of the Southern Kansas Railroad.

In 1873, when the Erie Railroad was reorganized, Mr. Chanute was made Chief Engineer of that Railroad. It was then proposed to double-track this road, to standardize its 6-foot gauge, to extend the line to New England and to Chicago, and generally expend about $50,000,000 in improvements. This promised well for the Engineer, but the financial panic of 1873 upset the arrangements made in England for funds. Less than $5,000,000 was expended on the Road, which served, however, to double track the main line, change the gauge to standard and to improve the gradients so that during the ten years which Mr. Chanute spent on the Erie the average freight train could be increased from 18 cars to 35 cars. For a time, he was in charge of the motive power of the Railroad so as to readjust the distribution of locomotives.

Upon his return to New York in 1873 he observed that "Rapid Transit" in that city had been under discussion for nearly twenty years without any definite results. The necessity for more rapid communication was evident; many projects for superseding horse cars had been proposed but none was recognized as solving the problem and the public seemed to be completely at sea. After some unsatisfactory inquiries as to the reasons for this condition of affairs, Mr. Chanute concluded that the question would be best settled by investigation through a committee of the American Society of Civil Engineers, and having been made Chairman of such a committee, consisting of W. N. Forney. Ashbel Welch, Chas. K. Graham, and Francis Collingwood, they undertook to collect all the data and facts. These were chiefly contained in pamphlets aggregating about 4,000 pages octavo, and after this literature had been digested communications were invited by circular. Five public meetings were held to give hearings to all who chose to appear, interviews were had with organized bodies and with citizens who were likely to possess information, and a canvass was made of property owners and tenants upon the proposed routes.

The resulting report was made public in 1875. It recommended substantially the plan subsequently carried out, i.e., the building of four lines of Elevated Railroads along the Avenues, to be operated by steam locomotives, and stated that such lines would be profitable, which was not generally believed by moneyed men at that time. The report was, at first, vigorously assailed by men interested in other projects, but the public accepted it and it was almost immediately followed by the requisite legislation and the building of the Roads. In this investigation, the laboring oars were wielded by Mr. Chanute and Mr. Forney, then Editor of the Railroad Gazette, and they wrote the reports and appendices. The former, indeed, had done all of his work at night, so as not to interfere with his Railroad duties, and he found himself so prostrated upon its completion that he had to take a vacation, and he went to Europe for four months to recruit.

In 1880, Mr. Chanute was appointed Chairman of a committee of the American Society of Civil Engineers to report upon wood preservation. The investigation occupied five years, and resulted in the publication of a report of great value, which was the authority on the subject for many years. In 1885, a number of American Railroads called on the Chairman to build plants for wood preservation, and he after that became interested in wood preservation as a business.

In 1883, he resigned from the Erie Railroad and removed to Kansas City, opening an office as Consulting Engineer. In this capacity he had charge of the design and construction of the iron bridges of the Chicago, Burlington & Northern Railroad, between Chicago and St. Paul, and later he performed similar work for the Atchison, Topeka & Santa Fe Railroad on the extension of its line from Kansas City to Chicago, involving, besides a number of minor streams, the bridging of the Missouri River at Sibley and of the Mississippi River at Fort Madison.

As long ago as 1874, Mr. Chanute had become interested in Aviation, but it was not until 1889, when he moved to Chicago and thenceforth made that city his home, that he could find the time to devote himself seriously to the solution of this "Problem of the Ages." With characteristic industry and thoroughness, he engaged in an extensive correspondence with men all over the world interested in this subject and gathered and systematized all information of importance which he could find, investigating the records of experiments of the past two or three hundred years. The fruit of these labors appeared in a series of articles entitled "Progress in Flying Machines." first published in the American Engineer and Railroad Journal, New York (October 1, 1891, and following issues), and republished in book form in 1894. This publication was of the greatest importance to the advancement of the art. for not only were all experiments of any importance described in detail and the principles elucidated, but the opinion of the author on the causes of the failure and the probable direction in which improvement might be expected, were stated.

About this time Otto Lilienthal had been making successful gliding experiments near Berlin, and these induced Mr. Chanute to build a Lilienthal glider and attempt experiments with man- carrying models in continuation of his previous experiments with small models. The site chosen was Dune Park, near the present town of Gary, on the sand dunes of Indiana. His purpose was mainly to attain equilibrium in the air, and he hoped to accomplish this by adjustments that should be largely automatic. Un- like unfortunate Lilienthal, who met his death in August 1896, while making one of his gliding experiments, Mr. Chanute's many experiments—upwards of 200 flights—were free from any misadventure to life or limb. The Lilienthal glider was an unwieldy monoplane requiring great skill in management. It was soon abandoned, and a multiplane glider substituted, and this in turn was replaced by the much simpler and more efficient biplane, the prototype of the present Wright aeroplane. These experiments were described by Mr. Chanute in a paper read October 20, 1897, before the Western Society of Engineers and published in its Journal.

All these experiments and investigations were made in a genuinely scientific spirit and at his own expense, free to all who were interested, and conducted without any thought of pecuniary or other benefit to himself. He had at an earlier time described his efforts with characteristic modesty as "Giving much of his leisure to the investigation of the chances of success in the possible solution of the problem of Aerial Navigation, not with the expectancy of solving it himself, for he held this would be the work of many men by a gradual process of evolution, but with the hope of advancing the question a little, and making the process easier for those who came after him, by eliminating some of the causes of past failures and laying down the principles which will have to be observed."

On October 20, 1909, just twelve years after the reading of his first paper on Aviation to the Western Society of Engineers, he read his last paper, entitled "Recent Progress in Aviation," in which he described the bewildering record of successful flights that had been achieved, giving a complete chronology of Aviation from December 17, 1903, when the Wrights made the first successful man-flight in history, to October, 1909.

After Mr. Chanute's death, the Aero Club of Washington said of him: *"Lilienthal, Chanute, Langley, and Maxim are the four names that will ever be inseparably linked with the early stages of flying-machine development, the stages that preceded the successful invention of the first man-carrying machine by the Wright Brothers. These four men elevated an inquiry, which for years had been classed with such absurdities as the finding of perpetual motion and the squaring of the circle, to the dignity of a legitimate engineering pursuit."*

Mr. Wilbur Wright in *Aeronautics* paid him the following tribute: *"If he had not lived, the entire history of progress in flying would have been other than it has been, for he encouraged not only the Wright Brothers to persevere in their experiments, but it was due to his missionary trip to France in 1903, that the Voisins, Bleriot, Farman, De Lagrange, and Archdeacon were led to undertake a revival of aviation studies in that country, after the failure of the efforts of Ader and the French government in 1897 had left everyone in idle despair. Although his experiments in automatic stability did not yield results which the world has yet been able to utilize, his labors had vast influence in bringing about the era of human flight. His 'double-deck' modification of the old Wenham and Stringfellow machines will influence flying machine design as long as flying machines are made. His writings were so lucid as to provide an intelligent understanding of the nature of the problems of flight to a vast number of persons who would probably never have given the matter study otherwise, and not only by published articles, but by personal correspondence and visitation, he inspired and encouraged to the limits of his ability all who were devoted to the work. His private correspondence with experimenters in all parts of the world was of great volume. No one was too humble to receive a share of his time. In patience and goodness of heart he has rarely been surpassed. Few men were more universally respected and loved."*

Mr. Chanute became a member of the American Society of Civil Engineers on February 19, 1868; served as a Director for four years, Vice-President for two years, was President in 1891, and for the five succeeding years was ex-officio a member of the Board of Directors.

He was elected an Active Member of the Western Society of Engineers July 12th, 1869; was President in 1901 and was elected an Honorary Member January 5, 1909. Some years ago, he presented to that Society a fund of $1,000.00, the interest of which was to provide bronze medals to be awarded annually for the best papers on Civil, Mechanical, and Electrical Engineering subjects. It is noteworthy that the committee to award prizes for papers read to the Society in 1909 reported, after Mr. Chanute's death, that his paper of October 20, 1909, on "Recent Progress in Aviation," was the first in merit of all the papers submitted during that year and the Society presented the medal to his family.

He was elected an Honorary Member of the British Institution of Civil Engineers May 21, 1895.

He was an Honorary Member of the Canadian Society of Civil Engineers, a Corresponding Member of the French Society of Civil Engineers and also of the Chilean Society of Engineers.

He was a Member of the American Railway Engineering Association, the American Institute of Mining Engineers, the Society for the Promotion of Engineering Education, and the American Association for the Advancement of Science.

He was a Fellow of the American Aeronautical Society, an Honorary Member of the Aero Club of America, and was President of the Illinois Aero Club. He was also honored by foreign Aero Clubs. The Aeronautical Society of Great Britain awarded him a gold medal in recognition of his distinguished services in promoting the art of aviation.

Mr. Chanute held the degree of Doctor of Engineering from the University of Illinois.

Mr. Chanute was Chairman of the Executive Committee of Engineering Societies which had charge of the International Engineering Congress at the World's Columbian Exhibition in Chicago in 1893.

He was a frequent contributor to the publications of Scientific Societies and to the various technical journals.

A list of the various contributions to Engineering literature, prepared by the deceased, begins with a paper on "Pneumatic Bridge Foundation," published in the *Journal of the Franklin Institute* in 1868, and ends with a paper on "The Present Status of Aerial Navigation." This last paper was published in *Science*, on December 29th, 1910. This list contains 60 titles, the two principal ones being his books on "The Kansas City Bridge" and "Progress in Flying Machines," which have been referred to.

In his relations to his fellow men, it is sufficient to record that he was unselfish, just, and kind. The present generation needs no eulogy of him, and for posterity it may be simply stated that he possessed the qualities of mind and heart which endeared him to his friends, caused those with whom he came in contact to respect and admire him, and which, ripened and chastened by the hard school of experience, produced in him one of our foremost Engineers, whose busy life was rich in works and achievement, and kindly and generous acts, commanding our admiration and prompting our love.

Mr. Chanute was traveling with his daughters in Europe in the Summer of 1910, when he was taken ill with pneumonia. For a time, his illness was of so serious a nature that a fatal termination was expected, but he recovered sufficiently to return to America, and, after a lingering illness, died at his home in Chicago, November 23, 1910. His remains were interred at his old home in Peoria, Illinois.

Mr. Chanute was married in 1857 to Miss Anne Riddell James of Peoria, Illinois, who died in 1902. He is survived by a son and three daughters.

~ Onward Bates, Robert W. Hunt, Charles F. Loweth, Charles L. Strobel

Some of Chanute's posthumous honors include:

- Induction into the National Aviation Hall of Fame.

- Induction into the International Aerospace Hall of Fame.

- Honored by four foreign civil engineer societies: British, French, Canadian and Chilean.

- A traffic island in Paris, France was named for him.

- Chanute Air Force Base in Rantoul, Illinois, was named for him in July 1917. The decommissioned base is now home to the "Octave Chanute Aerospace Museum."

- The U.S. Postal Service issued two airmail stamps in Chanute's honor in 1979. At least one stamp was the project of the Neosho Valley Historical Society to have a stamp commemorate the engineer for whom our city is named.

- Awarded a gold medal by The Aeronautical Society of Great Britain.

- The first American in outer space, Alan Shepard, wrote "…Chanute was a genius who might well be called the 'Father of Aviation'…."

- Chanute, Kan., placed a sculpture in 2002 to honor their town's namesake.

2020 KANSAS CITY BRIDGE ENGINEERING RETROSPECTIVE.

by Brian Snyder, PE

The Kansas City Bridge was, in its time, a critical project and an amazing feat by a gifted American engineer, Octave Chanute. But there is no historical landmark to mount a bronze plaque in his honor.

The chronicles of the creation of critical infrastructure of an expanding nation are rarely found in history books or discussed in classrooms. A few stories of monumental bridges in developed urban settings are written and celebrated by historian, politicians, and even Ken Burns on PBS programs. However, at the edge of the 19th century frontier in the small Town of Kansas, the bridge over the Missouri River would receive little attention nationwide in its day.

As what happens with many historic structures that have been demolished and removed from our cultural landscapes and collective memories, we rarely honor landmarks that are no longer visible. Their removal could be considered by some

Octave Chanute

as failures. The American Society of Civil Engineers (ASCE) does not list the Kansas City Bridge in its Civil Engineering Historic Landmark Program. And in ASCE's current (2020) list of the 46 great Civil Engineers of the world honored for "their creativity and ingenuity to lead the way to innovative civil engineering design," Octave Chanute is absent. Is the omission because this accomplishment no longer remains as a tribute for the civil engineering profession? Would Chanute rank in the top 50 great civil engineers by ASCE? Still fewer professionals get honors for mentoring and inspiring others who have achieved the highest honors and recognitions. Chanute catapulted the career of his engineering assistant, George Shattuck Morison (1843-1903). Together they published their final report of their 1869 victory over the Missouri River in a book titled, *The Kansas City Bridge*.

And, Orville and Wilbur Wright relied on Chanute's vision and consultation in developing the world's first flying machine.

My introduction to Octave Chanute was in 1994. Admittedly, the connection had less to do with engineering but with a valiant attempt to try and save the architectural integrity of Kansas City's Historic Garment District, a National Historic Registry District in the downtown central business core of the city. Developer DST, and Broadway Square Partners, expressed their plan to demolish a building—dedicated in the name of Octave Chanute—in order to build a small parking structure. The Chanute Building, constructed in 1906 and located at 407 West Eighth Street, was a 6-story commercial and warehouse building in the Neo-Classic Revival Style, and had the capacity to be adapted for reuse for multiple purposes including as a high-capacity parking garage. The nonprofit Historic Kansas City Foundation not only led the support for preserving the building but also held a preservation easement for the façade of the structure. The developer's representative suggested that the building was insignificant, uninteresting, and that even Chanute himself would be insulted to be associated with it. The developers won the battle and demolished the building leaving only a story and a half of the front façade with no purpose but to visually screen an unattractive parking structure. As for Octave Chanute, a well-known civil engineer who advanced the engineering profession working with practically every form of transportation in the country, a valuable building dedicated in his name was razed for the transportation and automobile parking requirements for a single business for that particular moment in recent history. And after 25 years, the parking demands of the Historic Garment District have driven well past the capacity of that short-sighted, small parking facility.

It is interesting (and ironic) that the very first bridge that crossed the Missouri River was over 200 miles upstream from the city of St. Louis, a community 75-years-older than the Town of Kansas with the resources and better assess to materials, machine shops, and labor to take on such a project. Even Jefferson City, the honored State Capital of Missouri, did not have a bridge. Most of the transcontinental rail traffic was heading west and southwest and the Missouri River was not an obstacle until it took a sharp bend north. This left the old steamboat port of western Missouri as a prime location for the first crossing. The Eads Bridge crossing the Mississippi River in St. Louis was completed in 1874, five years *after* the opening of the Kansas City Bridge. Five years is a lifetime in the advancements in iron and steel designs and fabrication. The Mississippi River was considered a much more manageable crossing with predictable water levels, less turbulent currents, more stable navigation channels, and fewer tree snags floating downstream. There was also an extra 75 years of observations and data for water levels and flood events, ignoring Native-American folklore. In the most visible contrast to the Kansas City Bridge, the Eads Bridge—a Carnegie/Keystone Bridge Company structure as was Hannibal Bridge—is still standing and is now a celebrated image at the St. Louis waterfront with honors to its designer, James Buchanan Eads, and apparent naming-rights. And his bridge is further honored by the U.S. Department of the Interior as a National Historic Landmark and honored by the civil engineering profession through ASCE as a Historical Civil Engineering Landmark. But not all bridges from that time period were considered success stories.

Still fresh in the minds of Missouri officials who began in 1857 discussing the Kansas City Bridge with planners and railroad executives, was the first major deadly bridge collapse in American history which occurred just 40 miles east of Jefferson City, Missouri near Hermann. It happened mid-day, November 1, 1855, in a failed attempt of the newly charted Pacific Railroad to claim the first transcontinental railroad route to the Pacific Ocean. This proposed rail route had promise crossing the State of Missouri without having to deal with a difficult Missouri River crossing. A maiden excursion ride and celebration of the route's completion to Jefferson City departed St. Louis at 9 o'clock in the morning after much fanfare and speeches. Unfortunately, one of the spans of the bridge at the Gasconade River had not been completed. Temporary scaffolding which included trestlework was put in place just for the event. After the steam locomotive pulling ten full passenger cars and a baggage car traveling at 15 miles per hour, entered the 760-foot-long, multi-span bridge. Almost making it across the first span, the bridge structure started collapsing pulling all but one rail car 30 feet down into in the Gasconade River valley. Of the 600 passengers on board, 31 died including the Chief Engineer, Thomas S. O'Sullivan, along with many other railroad employees and city leaders from St. Louis. Two hundred passengers were injured. The loss for St. Louis was so great, the entire city was shut down to hold the funerals. The Railroad Company appointed a Commission to investigate the collapse and came to the conclusion that the high rate of speed was to blame for the accident. The recommended speed was only 5 miles per hour. But eyewitnesses had different conclusions and participated in a more objective Minority Report stating that the cause of the accident was under-sized structural members, insufficient foundations, and poor management and supervision of the construction. A month after the accident, George Vose, a graduate of the Lawrence Scientific School of Harvard and experienced railroad professional, published a scathing assessment of conditions surrounding the accident in a national railroad journal. He praised the Minority Report but added harsh criticism towards railroad management and the railroad industry itself which existed and functioned with practically no governmental oversight. Vose stated, "A lawyer must be admitted to the bar before his can practice, and physician must have a medical examination; but a man who is trusted with hundreds of human lives daily needs only a brazen face and plenty of influence to be Chief Engineer Roadmaster or Superintendent."

It would take another 52 years before the first engineer would receive an official engineering license from a state authority. While noting Missouri landmark events in structural collapses, we cannot leave this subject without mentioning the Kansas City Hyatt Hotel Skywalk collapse in 1981. Both tragedies had the human tragedy, the drama, and the contentious aftermath of determining causes and responsibilities. Both events led to improvements in the industry for the technical aspects of design and construction, but more importantly, improvements in integrating the design and construction processes resulting in better/safer projects.

Joseph Tomlinson III, Chanute's co-designer of the Kansas City Bridge superstructure, had just come from the Ashtabula River Railroad Bridge project in northeast Ohio after being fired for refusing to make design changes in the bridge's framework that he evaluated to be unsafe. Tomlinson's version of the truss designs had included both wood and iron elements, similar to that proposed for the fixed-spans of the Kansas City Bridge trusses. But Tomlinson worked for Amasa Stone, a railroad official, who demanded that the support trusses be all-iron. Interestingly, Stone actually owned the U.S. Patent for the Howe Truss which he purchased from his brother-in-law, William Howe, and used a variation of that truss design for the Ashtabula River Bridge. Amasa Stone ended up replacing all wood elements with iron without changing Tomlinson's overall design. When the erection of the Ashtabula River iron bridge was completed in 1865, it required immediate structural repairs and reinforcement which were performed in ways that made the bridge even more dangerous. The bridge would eventually collapse in the early evening of December 29, 1876, loaded with two locomotives and 11 passenger cars, and falling 150 feet down into the ice-clogged river. Oil lamps and kerosene heaters caused the rail cars to burst into flames resulting in the loss of 92 lives and injuries to 64 survivors.

Between 1850 and 1875, bridge failures were actually common, with at least 25 occurring each year in the country.

As the construction industries put further demands for more iron and steel products to extend bridges further and erect structures taller, it became critical to establish material and design standards. In the mid-19th century, each foundry would develop their own shapes for structural members, their own material testing procedures, and their own design standards and allowable loads for the use of their products. With the primitive analytic procedures where ultimate failure loads were compared to the maximum anticipated service loads, the ratio of these two loads is considered the "safety factor." For today's steel structural designed with complex 3-D computer modeling and steel industry metallurgical and design standards, in general, the safety factor is close to 2, sometimes more, sometimes less depending on the type of failure and the critical nature of the failure. In the early years of cast iron, a common range of safety factors from 6 to 10 was necessary to account for imperfections in iron casting such as unequal thicknesses and air holes embedded in the castings. Iron structural elements were also prone to buckle under compression forces, especially with incidental lateral loads applied.

Also keep in mind, in the 1860s we knew less about the lateral forces of wind and little about the effects of seismic events, especially as they related to specific geographic regions of the country. The American Society of Civil Engineers (ASCE) was founded in 1852 as the nation's earliest engineering organization setting out to facilitate the dissemination of technical information for the planning and design of buildings and critical infrastructure for a growing continent. Chanute was an early participant in ASCE. The American Society for Testing and Materials (ASTM) was established in 1898 in response to the railroad industries' failures of its steel rails. ASTM would eventually be the organization for material standards in practically all industries. The formation of the American Iron and Steel Institute (AISI) in 1908 followed by the American Institute for

Steel Construction (AISC) in 1921 further advanced standards in the design of steel structures.

After the Eads Bridge completion in 1874, the Brooklyn Bridge would open in 1883. It would not be until 1885 when the country would see its first steel-framed multistory building to reach 10 stories tall in Chicago, Illinois, the Home Insurance Building. It was eventually demolished in 1931 which ended up having a similar life span as the Kansas City Bridge. The life span is always critical with planning and designing bridge spans that bring life to communities.

And back in Chanute's birthplace, Paris, France, the monumental Eiffel Tower would open in 1889 symbolizing the country's dominance in the use of steel, its advances in structural engineering, and its prominence in style and architecture for the 19th century. The project also honored the engineer, Gustave Eiffel, who designed and constructed the tower which became the tallest man-made structure in the world at that time.

The Kansas City Bridge represented one of the more substantial steel-framed structures on the continent for its time and especially for this region. The use of iron building products this far west had been limited to secondary structural members (i.e., hardware, anchor straps, iron rod ties, and wall washers to brace masonry structures).

Chanute's bridge superstructure design incorporated a unique combination of wood, cast iron, and rod iron. Tomlinson had likely contributed to this design after the debacle at the Ashtabula River Bridge, created by his past employer, in working with an all-iron bridge.

In early August 1867, after 6 months into the construction of the piers, invitations were sent out requesting proposals for the construction of the superstructures of the Kansas City Bridge. Preliminary designs had been prepared by Chanute and Tomlinson for the stationary spans which were to be assembled in iron and wood. In general, wood was used for structural members in compression. Iron was used for structural members in tension. For the 363-foot-long pivoting span, the bidders were asked to perform the design and the fabrication for this more complicated portion of the bridge due to scheduling constraints and to take advantage of the bidder's expertise and experience in these special structures and mechanisms. The pivoting span was designed and fabricated from cast iron and rod iron. For the stationary spans, Chanute encouraged the bidders to submit alternate designs that would be considered that would have advantages and cost savings over their designs. Chanute received 9 proposals from 5 bridge builders and accepted the bid from the Keystone Bridge Company in Pittsburg, Pennsylvania. A young entrepreneur named Andrew Carnegie founded the Keystone Bridge Company in 1865. The Kansas City Bridge was a signature project for his company.

The Keystone Bridge Company had the opportunity to submit acceptable alternate proposals for all-iron bridge trusses for the fixed-spans which would have increased profits for the company and allowed them more control of the fabrication and assembly processes. But that did not happen. This further reinforces the design proposed by Chanute and Tomlinson to use both iron and timber for the fixed spans for the sake of safety and reliability.

The other structural material was native timber where there was practically an unlimited supply. The lumber of choice was oak, elm, and sycamore likely harvested from old growth forests that surrounded the region. The timber was used for permanent and temporary foundation piles for supporting structures in and on the water to facilitate the construction of the masonry bridge piers. The deeper masonry bridge piers on the north side of the river where bedrock was lower were founded on permanent piles that were driven down to bedrock. The term "pile" refers to a long shaft driven into the ground to support structures on soft soil or driven down to bedrock for support. For this project, the piles were made from timber and would have resembled large utility poles. For the support of a large bridge pier, it would take up to 144 piles for an adequate base of a single pier foundation. Since the permanent foundation piles were well below the water line in consistent environmental conditions, the old growth timber piles could last for hundreds of years. Timber structural elements used in the bridge deck and truss members above the river were not only exposed to varying stresses and strains from the rail loads, but they were also constantly exposed to temperature and moisture changes, freezing/thawing, and microbial, fungi, and insect attacks. From my own experience in the restoration of timber structures from the mid-19th-century, the old growth timber has higher density, durability, and stability properties with at least twice the strength and longevity of what is available today. In many cases, you can barely drive a nail it. Screws typically break off when using an electric screwdriver. The worse enemy to timber is moisture. Chanute incorporated the best protective coatings that were available in that time. Unfortunately, coatings can inhibit natural drying of the wood which can also accelerate deterioration. And successive recoating of the timber can hide deterioration from routine inspections. Chanute and Morison mention in *The Kansas City Bridge* that he anticipated wood elements of the bridge superstructure could be replaced with steel elements in the future. The wood elements would eventually suffer deterioration while advancements in steel quality standards could have advantages for new components. As a railroad consultant, Chanute would later study the longevity of timber railroad ties and structural elements which led to two U.S. Patents in timber preservation: Patent No. 430,068 on June 10, 1890, and Patent No. 688,932 on December 17, 1901.

Most of the timber and lumber were necessary for temporary structures to facilitate the construction of the piers over the water. Large wood caissons, which are watertight chambers, were constructed and lowered into the water to provide space, after dredging and pumping the sediment and water out, to place concrete fill as a level base to begin the massive stone masonry piers. Wood platforms had to be constructed on top of the piles and framing to provide access for worker and machinery. Large temporary caissons filled with sand were secured to the bottom of the river just to dissipate the strong river current and allow the work downstream for the pier construction caissons to remain stable and safe. And to facilitate the second attempt to construct Pier 4, a large wood-framed Foundation Shop that would have resembled a 12,000 square-foot, three-story mansion towering 50 feet above the water at its roof peak was constructed near the north shoreline of the river. In the peak of the construction of multiple piers, this portion of the river would have resembled a floating city or at least a huge traffic jam of steamboats. The use of timber was amplified due to the unavailability of a metal fabrication shop that could make the pieces necessary and in a timely manner. Chanute had to make major adjustments due to ever changing conditions on the Missouri River and timber was essentially the only structural material available.

The word "concrete" is never used in *The Kansas City Bridge* book. Instead, Chanute, a French native, uses a French translation, "beton." This leads to confusion when reading, even for the technically trained. The concrete mixed for the bedding and leveling of the pier bases was likely proportioned from lime, sand, and crushed rock. Portland Cement had not yet made it to this part of North America. "Caisson" is also French.

MODERN EXAMPLES

OF

ROAD AND RAILWAY BRIDGES;

ILLUSTRATING

THE MOST RECENT PRACTICE OF LEADING ENGINEERS

IN

EUROPE AND AMERICA.

BY

WILLIAM H. MAW AND JAMES DREDGE.

PARTIALLY REPRINTED FROM "ENGINEERING."

"Bridge Over the River Missouri at Kansas City,"
is a featured chapter in this 1872 compendium.

LONDON:
PUBLISHED AT THE OFFICES OF "ENGINEERING," 37, BEDFORD STREET, STRAND.
BERLIN: MESSRS. A. ASHER & CO., 11, UNTER DEN LINDEN.
1872.

Chanute's presentation in Chapter 3 of foundation Piers 1 through 7 spared no details with his descriptions of components and systems together with his artfully crafted drawings. It is clear that construction of Pier 4, in particular, was the primary inspiration for publishing the book. First, it is noted that one third of the main chapters of the book and half of the plate images are dedicated to all 7 of the bridge pier foundations. And half of those pages covered just the construction of Pier 4, a 20-month-long process that consumed forty percent of the foundation budget. Keep in mind, all the foundation piers took 25 months to construct with most of that work scheduled between August and December, or what Chanute referred as "the season between two floods," when the water levels and currents made the work more manageable.

The story of Pier 4 would begin in September 1867 with a similar scenario as all the rest of the piers, driving temporary timber piles to prepare a framework for which to work from. But delays in the work due to large deposits of loose sand and the work on the adjacent piers taking away the pumping and dredging equipment pushed the work into the winter months. By mid-February, they were approaching the wet season and apparently decided to move forward setting up again the heavy steam-powered dredging equipment above the caisson. The river level began to rise quickly on March 8, 1868, while the dredging was in progress. By March 17, ironically St. Patrick's Day, their luck ran out. The heavy river current was much stronger on the south side of the caisson versus the north side closer to the riverbank. This led to uneven scouring (erosion) of the riverbed supporting the caisson. It did not help that the caisson assembly was top-heavy with the mounting of the dredging equipment. The caisson began to tilt which, by the following morning, left only one corner of the caisson protruding above the water. By 2 p.m., the entire caisson literally disappeared just hours after the dredging equipment was quickly removed. The loss of the caisson and all the work associated with it was bad enough, but it also left the site on the riverbed with the wreckage of timber, sand, and stone. After much deliberation, it was decided to move the location for Pier 4 to the south 50 feet which was actually into deeper water and stronger currents. The 50 feet allowed the new pier site to safely clear the wreckage with minimal impact to the bridge trusses which were at least in the design phase and preparation for fabrication. The two adjacent spans of 250 and 200 feet to be supported from Pier 4 could just switch places. In every historic photo and sketch you see of the Kansas City Bridge, this 250-foot span, that is surrounded by the three shorter spans that make up the northern portion of the bridge, is noticeably taller.

Reconstruction of the new Pier 4 caisson would not begin until June 25, 1868, and by early August, work began in driving 60 timber piles into the riverbed that would support the three-level Foundation Shop. This temporary timber-framed structure would have platforms to access steam engines, four dredging machines, pumps, and material handling equipment. Chanute devised a method for which the construction of the massive stone masonry pier could start above the water and be suspended from the framework while incrementally lowering it down into the water and while dredging, pumping, and displacing the riverbed material and eventually exposing bedrock. Levels of the stone masonry work would progress incrementally until the pier was safely resting and secured to bedrock. This innovative operation required the conventional dredging equipment to be modified for a fixed mounting to the Foundation Shop framing while making the dredging frame assembly adjustable (telescopic) to be able to increase the depth of the dredging as silt and mud was removed from the riverbed.

This massive undertaking explains why the Foundation Shop to facilitate this process had to be so large. The pier caisson measured 70 feet by 22 feet in plan dimensions which required the Foundation Shop to be approximately 95 feet by 65 feet, four times the pier caisson footprint area. The shop was enclosed with siding and a double-pitched roof allowing shelter for the workers since this work was extended into the winter months. With all steam systems and equipment running, it likely radiated heat for the workers. The operations and construction of the pier ended up going very smoothly. They were able to work 24 hours a day while catching up to their original schedules after lost time from the first attempt and failure.

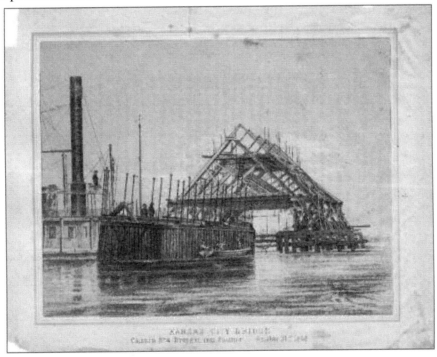

The completion of Pier 4 (caisson setting underway in the above engraving) ended up being an overall success. In fact, Chanute referred the construction of Pier 4 as "the most successful" of all the piers. This is what was so extraordinary for this huge undertaking. Chanute and his team not only had to design and build a bridge. They also had to design, fabricate, assemble, operate, maintain, and decommission unique construction equipment and support structures for 7 different bridge piers, each with its own unique, unforeseen, and changing conditions with limited resources on the edge of the frontier. In a footnote below the conclusions of the Pier 4 narrative, Chanute noted that, "a patent for this method of founding has been applied for by the authors of this volume." The following is a reproduction of Chanute and Morison's original patent No. 98,848, approved by the U.S. Patent Office on January 18, 1870, for their "improved dredging machine:"

UNITED STATES PATENT OFFICE.

O. CHANUTE AND G. S. MORISON, OF KANSAS CITY, MISSOURI.

IMPROVED DREDGING-MACHINE.

Specification forming part of Letters Patent No. **98,848,** dated January 18, 1870.

To all whom it may concern:

Be it known that we, OCTAVE CHANUTE and GEORGE S. MORISON, of the city of Kansas, in the county of Jackson and State of Missouri, have invented a new and useful Improvement in Dredging-Machines, the object of which improvement is to increase the range of the machine, thereby enabling the same machine to work at greatly-varying depths.

The nature of this invention consists in so mounting an endless-chain dredging-machine that the frame carrying the chain and buckets may be lengthened to suit an increased depth of excavation, and also be raised and lowered independently of such lengthening, to follow the varying levels of the material excavated, as will appear more fully from the following specification and accompanying drawings.

The frame carrying the chain and buckets is of two parts—a double outer frame, A A, and a single inner frame, B B, movable within the former—the two being held together by the bolts $a\,a\,a\,a$. The lower tumbler, E, is mounted at the foot of the inner frame, and the upper tumbler, F, on the head-blocks C C, also carry the counter-shaft i. These head-blocks are carried by the screws $b\,b$, set in the head of the double frame A A, and passing through the head-blocks, which may be raised or lowered by the adjusting-nuts above and below. The bucket-frame is free to move vertically in the stationary frame D D, and may be raised and lowered by the chains $d\,d$ passing through the sheaves $e\,e$ on each side of the double frame A A. The length of the bucket-frame may be increased by adding additional lengths to the bucket-chain, withdrawing the bolts $a\,a\,a\,a$, and raising the double frame A A upon the single frame B B till the desired length is obtained. The two frames should then be reunited by replacing the bolts $a\,a\,a\,a$, and any slack remaining in the bucket-chains taken up by the adjusting-screws $b\,b$. The frames A A and B B should be bored to receive the bolts $a\,a\,a\,a$, the distance from hole to hole being one-half the distance from bucket to bucket, or, in a machine like that shown in the drawing, three times the length of each link of the bucket chain. The driving-power is communicated by the belt $f\,f$, which turns the upper tumbler, F, by means of the pinion g, working on the gear-wheel h, the adhesion of the belt being secured by the weight of the tightener H, sliding in a vertical frame. The buckets are made with curved or inclined bottoms, as shown on the drawing, so that the form of the inverted bucket shall throw forward the excavated material falling from the bucket above, thereby dispensing with the rockers usually needed to shed the sand from vertical endless-chain dredges.

This improvement may be applied to any dredging-machine of the endless-chain pattern; but it is especially used in excavating for deep sand foundations, and wherever excavations of great depth are to be made within a limited area.

The machine should be so mounted that at the beginning of operations the buckets will scrape the sand when the bucket-frame, set at its shortest length, is raised as high as convenience allows. As the excavation proceeds the bucket-frame should be fed down by the chains $d\,d$ as far as the driving-gear will admit. An additional length of bucket-chain should then be added, the bucket-frame lengthened by raising the outer frame, A A, and the dredging continued as before, the same process being repeated until the excavation is carried to the desired depth.

For very deep work the bucket-frame may be made of three or more parts sliding within one another, in place of two; or the combination of the two motions may be obtained by some other appliance; and the frame may be mounted on an incline, instead of vertically, if preferred.

A system of gear-wheels may be substituted for the driving-belt and tightener, and the bucket-frame may be raised and lowered by screws or a pinion and ratchet in place of the chains $d\,d$ and the sheaves.

What is claimed in this invention is—

1. The combination of the two motions—of the motion within the bucket-frame, by which the distance between the tumblers is varied to suit the depth of excavation, and of the motion of that frame independently of this change of length, by which the buckets are fed into the material excavated.

2. The compound or telescopic bucket-frame, of two or more parts, sliding within one another.

3. The adjustable head-blocks, regulated by screws, substantially as described.

O. CHANUTE,
G. S. MORISON.

Witnesses:
GEORGE E. PITKIN,
H. C. BRYANT.

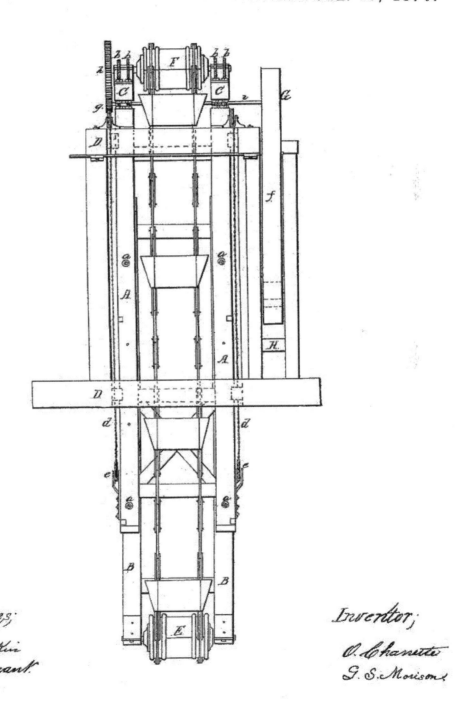

CHANUTE & G. S. MORISON.
DREDGING MACHINE.

No. 98,848.

Patented Jan. 18, 1870.

Witnesses;

Geo. C. Pitkin
H. C. Bryant

Inventor;

O. Chanute
G. S. Morison

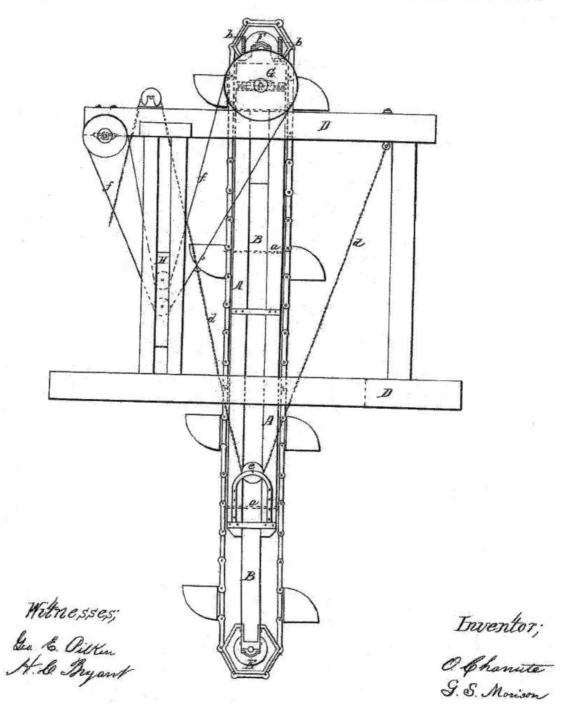

There is much detail in each of the chapters Chanute and Morison's *The Kansas City Bridge* book. It presents data for testing materials, the weights and calculated deflections of each of the trusses, lists of all equipment and materials used, the results of structural calculations, the loads on each of the bridge piers, the stress and stains on the steel work, a breakdown of construction costs which total $1,093,177.58, unit pricing on all materials, traffic data on the bridge's first 7 months of operation, data from measurements and instruments on the installation of Pier 4, and readings of water levels during the 3 years of construction.

The Missouri River became not just the challenge to span over but also played the part as the adversary who worked against them at every step, and in incalculable ways.

On the title page of Chanute and Morison's book, the subtitle references "the Regimen" of the Missouri River which I interpret to mean the ultimate "authority" reminding the readers who was really in charge of the operation. The river was the judge and jury for the construction of the foundations. And my favorite description of the Missouri River mentioned in Chapter 2, as a local saying, that the river, "has a standing mortgage on the entire bottom land from bluff to bluff, and the farmer on the Missouri bottom often learns to his sorrow, by the loss of his farm, that real estate is not always immovable property." Chanute further personifies the character of the river as having moods that "were constantly changing."

After successfully completing the Kansas City Bridge, the Keystone Bridge Company would construct another signature project, the Eads Bridge in St. Louis, Missouri. The company's owner, Andrew Carnegie, would eventually lead the expansion of the country's steel industry into the 20th century becoming the richest man and the most generous philanthropist in America. In 1891, he constructed Carnegie Hall in New York City making it still today (2020) the premiere performing arts venue in the world. Exactly 100 years after the opening of the Kansas City Bridge, Carnegie Hall hosted one of the first live performances of the song, "Bridge Over Troubled Waters" written by Paul Simon and performed by Simon and Garfunkel.

Octave Chanute's assistant engineer, protégé, and co-author of their 1870 book, *The Kansas City Bridge*, George S. Morison, performed critical engineering duties. After the completion of the Kansas City Bridge, Morison achieved his own notoriety on landmark bridge and civil engineering projects. He would go on to serve as one of the leading authorities for the U.S. government's commission on locating the route for the Panama Canal. In 1895, Morison served as president of the American Society for Civil Engineers and has since been honored by ASCE on their list of Historic Civil Engineers. Chanute's roll as chief engineer and supervisor/mentor to Morison led to great advancements in the civil engineering profession.

In his later years as an engineering consultant, Octave Chanute would provide engineering support for Orville and Wilbur Wright in the development of the world's first flying machine. In addition to three aviation patents in the United Kingdom, Chanute's work would result in four more patents from the U.S. Patent Office:

Patent No. 582,718 on May 18, 1897, for a Soaring Machine;

Patent No. 582,757 Means for Aerial Flight

    (on behalf of Louis Mouillard, with one half being assigned to Chanute);

Patent No. 606,187 Soaring Machine (with William Paul Butusov; Chanute collaborated and paid for the patent process and was assigned one half) on June 28, 1898; and,

Patent No. 834,658 October 30, 1906, Means for Aerial Flight (or, glider launcher).

The Kansas City Bridge book is a comprehensive account of the planning, design, and construction of the Kansas City Bridge over a river described in the very first sentence in the book, "as almost incapable of being bridged." Octave Chanute and his team were successful in taking on the challenge. The highlight of the book was not what was visible above the water surface but was buried well below into the bed of "that river" at Pier 4. The innovative design of the construction method that was necessary to construct this pier, as presented and introduced in the book, appears to be the book's primary mission. The publication also provided a unique opportunity to present the entire landmark project as an educational tool for the civil engineering community.

Chanute's book is niche and not for every reader. I do recommend at least skimming the 140 pages and examining the sketches to get an idea the level of detail the design and construction team went through to accomplish this huge project in this isolated location on the frontier on "that river."

For the average reader beyond the civil engineering community, it is hoped that this 150th Commemorative Edition provides expanded content and scholarship of interest to the general Kansas City history aficionado.

ABOUT THE AUTHORS.

Octave Chanute – Biography appears above in Volume 1.

George S. Morison – Biographical sketch appears in Volume 2.

Bill Nicks, Jr., is the retired Director of the Lenexa Kansas Parks & Recreation Department. Nicks became interested in Chanute when he discovered how Lenexa was significantly impacted by this great man. Nicks then learned about Chanute's Kansas City Bridge and soon discovered that the Wright Brothers, too, had been influenced by Chanute's vision.

Believing Chanute to be under-appreciated, Nick studied Chanute intensively and began in 1995 performing for the public first-person, living history re-enactments impersonating Octave Chanute.

Brian Snyder is a professional civil/structural engineer registered in the State of Missouri and retired from Burns & McDonnell after 34 years of service. He received his BS & MS in Civil/Structural Engineering from the University of Missouri–Rolla in 1980 and 1982.

Snyder served as a Board Member of the Jackson County (Mo.) Historical Society and Chair of the Independence Heritage Commission. He was a volunteer consultant for the relocation/restoration of the circa 1879 Chicago & Alton Railroad Depot in Independence and Charter Board Member of the Friends of the C&A Depot. His memberships also include the American Society of Civil Engineers, Historic Kansas City Foundation, Missouri Preservation, and the Association for Preservation Technology. Snyder is currently (2020) serving on the Board of Trustees of the State Historical Society of Missouri.

Snyder and his wife, Sharon, are currently working on the archaeology, stabilization, and restoration of the circa 1840/1852 Owens-McCoy House (owensmccoyhouse.com), a National Park Service Certified Santa Fe Trail Property. William McCoy who had just completed his first (and only) one-year term as Mayor of the newly incorporated City of Independence, purchased the 14-acre property and 10-year-old house in 1850 from the estate of Samuel Combs Owens, the first Jackson County Clerk in 1827, and a prominent Santa Fe Trail merchant. Owens was also one of the original incorporators who founded the Town of Kansas, the village with a front row seat for the construction and opening ceremonies of Octave Chanute's Kansas City Bridge.

V O L U M E   1   N O T E S.

[1] Glaab, Charles N. *History of the Railroads in Kansas City: Community Policy in the Growth of a Regional Metropolis*. (Madison, Wi.: State Historical Society of Wisconsin, 1963).

[2] Chanute, Octave Alexandre, and George S. Morison. *The Kansas City Bridge: With an Account of the Regimen of the Missouri River, and a Description of Methods Used for Founding in that River*. (New York, Ny.: D. Van Nostrand, 1870). The original book measures 12" x 9.5" and is covered in brown cloth with stamped gilt title. There are 140 pages of text on history, engineering, technology, and one appendix that includes the names of management-level employees in charge of building the bridge. In the back of the book, there are 12 detailed, foldout plates measuring 16" x 11" showing details of the Kansas City Bridge construction. The copy from which this reprint is derived was originally a monograph in the Washington University Library in St. Louis, Missouri. Engineer Brian Snyder procured the book for sale online. After scanning the book, he donated the original to the Jackson County (Mo.) Historical Society Archives, Independence, Missouri. Snyder shared his scans with *Engineered Irony* author David W. Jackson who manipulated the images to standardize and clean up blemishes as much as possible before re-inserting into book format. The large plates were resized to fit the open page spread of this 8" x 10" format. Recognizing this is not ideal, it is the best solution settled on for this 150th Commemorative Edition. Scans of the book, including elevations on 13 plates, may be found on various online, digitized portals. Another copy consulted in the production of Engineered Irony was Chanute's personal library copy of *The Kansas City Bridge.*

[3] Montgomery, Rick. "The Bridge: A Footnote: Opening Day," *Kansas City* (Mo.) *Star*, 26 Oct 1997. The *Daily Journal of Commerce* (Kansas City, Mo.), 7 Jan 1868, also syndicated the Omaha Herald stating that "the first loaded train of cars passed over the Union Pacific Railroad bridge which now spans the Missouri River, and this was the first train of cars that ever crossed that monarch of rivers, either at this point or any other, from its mouth to its source." Union Pacific Historical Museum curator Paul Rigdon wrote to Kansas City historian James Anderson on February 1, 1950, and verified that the "pile bridge, driven through holes cut in the ice, to provide a means of crossing during the winter months when the river was frozen over all or part of the time and when the operation of the 'transfer boats' from shore to shore was impeded or prevented by the ice. It's use was not confined, however, to the movement of construction materials: regular freight and passenger traffic was moved across the pile or 'ice' bridge in both directions. We have in our historical collection several photographs of trains of cars being taken across the Missouri River on one of these ice bridges. When the ice broke up and the river cleared in the spring, the bridging of course, except the deck and such of the bents as could be salvaged, was washed away....." He continued there may have been "five of these temporary bridges driven through the ice, and used during the winters of 1866-67 to 1871-72, before the first permanent bridge across the Missouri at Omaha was completed in March 1872. From the standpoint of permanency, the iron bridge constructed over the Missouri, completed in 1869, near Kansas City, by the Hannibal & St. Joseph Railroad Company, was the first bridge across the river." Letter at the State Historical Society of Missouri—Kansas City Research Center, Native Sons and Daughters of Greater Kansas City papers, K0395, Box 12, Folders 13-18. In this same folder is a letter to Anderson from Donald Ashton, Burlington Lines, who concludes his 3-page letter, "It is even questionable in my mind whether the temporary bridges at Omaha can be considered a bridge."

[4] http://www.lib.uchicago.edu/e/crerar/exhibits/chanute3.html (viewed 1 Nov 2020).

[5] Montgomery, Rick. "Desperate for a break, the city politicked and pleaded and got The Bridge," *Kansas City* (Mo.) *Star*, 12 Oct. 1997.

[6] Eight days is a Wikipedia estimate. Simine Short wrote, "Enterprising travelers could now journey from New York to San Francisco on various trains in six days, eight hours, assuming there were no unforeseen difficulties...." Short, Simine. *Locomotive to Aeromotive: Octave Chanute and the Transportation Revolution*. (Urbana, Il.: University of Illinois Press, 2011), 50.

[7] Short notes on page 52 that, "Chanute ordered 250 copies for the bridge company, including ten leather-bound books with fourteen photos of the construction." *Van Nostrand's Eclectic Engineering Magazine* reviewed Chanute and Morison's book; the review was syndicated in *Railway Times* (3 Feb 1872) 24:5:39. Also, Griggs, Frank, Jr., Ph.D. "Great Achievements: Notable Structural Engineers: Octave Chanute," *Structure Magazine* (March 2007) 58-59.

[8] "Flights Before the Wrights. Octave Chanute, Chicago. Aeronautical Pioneer, Engineer & Teacher. Online exhibition from 1 Nov 2001. (https://www.lib.uchicago.edu/collex/exhibits/flight-wrights/ viewed 31 Oct 2020).

[9] Also, *Kansas City Directory*, 1867-68, 14, 18, 55.

[10] https://kchistory.org/blog/hannibal-bridge-keeping-it-rail-1869 (viewed 24 Nov 2020).

[11] "The Semi-Centennial of the 'Hannibal' Bridge," *Kansas City* (Mo.) *Star*, 4 July 1919.

[12] "Kansas City's 'Immortal Bridge,' Still Serves March of Progress," *Kansas City* (Mo.) *Times*, 12 Aug 1940.

[13] First National Bank advertisement, *Daily Journal of Commerce* (Kansas City, Mo.), 18 Jan 1873, p. 3.

[14] "The West Kansas City Land Company Minute book, 1867-1894," 33-41, 151-152, as found in the Native Sons and Daughters of Greater Kansas City Collection, State Historical Society of Missouri-Kansas City by Louis Potts, Ph.D., and George F. W. Hauck, Ph.D., and published in "Frontier Bridge Building: The Hannibal Bridge of Kansas City, 1867-1869," Missouri Historical Review (Jan 1995): 89: 2: 146.

[15] "The New Air Ship Tested," *Kansas City* (Mo.) *Star*, 20 Sept 1896.

[16] https://lenexahistoricalsociety.org/origins-of-lenexa-2/ (viewed 20 Dec 2020). Also, Recorder of Deeds, Johnson County, Kansas: Deed: May 25, 1869. C.A. Bradshaw to Octave Chanute; Deed: Aug. 26, 1869. O. Chanute to H.M. Holden, et al.

[17] "Hannibal Bridge" (http://en.wikipedia.org/wiki/Hannibal_Bridge viewed 13 Oct 2009).

[18] Burnes, Brian. "The bridge that transformed Kansas City's future," *Kansas City* (Mo.) *Star*, 10 Oct 1994.

[19] "Kansas City's 'Immortal Bridge,' Still Serves March of Progress," *Kansas City* (Mo.) *Times*, 12 Aug 1940.

[20] "He Built the First Bridge," *Kansas City* (Mo.) *Star,* 29 July 1895, 2.

[21] Burnes, Brian. "The bridge that transformed Kansas City's future," *Kansas City* (Mo.) *Star*, 10 Oct 1994.

[22] "Built the Hannibal Bridge," *Kansas City* (Mo.) *Star,* 24 Nov 1910.